Genetic Diseases
or Conditions

Genetic Diseases or Conditions

Cystic Fibrosis, the Salty Kiss

Todd T. Eckdahl

MOMENTUM PRESS
HEALTH

Genetic Diseases or Conditions: Cystic Fibrosis, the Salty Kiss
Copyright © Momentum Press®, LLC, 2016

First published in 2016 by
Momentum Press®, LLC
222 East 46th Street, New York, NY 10017
www.momentumpress.net

ISBN-13: 978-1-94474-955-2 (print)
ISBN-13: 978-1-94474-956-9 (e-book)

Momentum Press Human Diseases and Conditions Collection

Cover and interior design by S4Carlisle Publishing Services Private Ltd., Chennai, India

First edition: 2016

10 9 8 7 6 5 4 3 2 1

Printed in the United States of America

Abstract

Cystic fibrosis (CF) is one of the most common genetic diseases, affecting about 70,000 people throughout the world, with over 1,000 new cases diagnosed each year. This book describes the symptoms of CF including lung disease, digestive problems, pancreatic insufficiency, liver disease, intestinal obstruction, and infertility. It explains how CF is caused by mutations in the *CFTR* gene encoding a protein ion channel that maintains the balance of salt and water in the lungs and other organs. The book presents CF as an autosomal recessive disease that can arise in families with no prior history of CF. The reader of this book will learn about treatments and therapies for CF, including antibiotics for infections, medicines for improved digestion, respiratory therapy, and pancreatic enzyme replacement. The book describes promising new pharmaceutical discoveries that enable personalized medicine for the treatment of CF. It evaluates the prospects for curing CF through gene therapy and explains how genome editing may be used in the future to correct the *CFTR* gene mutations underlying CF.

Keywords

autosomal recessive, *CFTR* gene, cystic fibrosis, genetic disease, liver disease, lung disease, mucoviscidosis, pancreatic disease

Contents

Acknowledgments .... *ix*

Introduction .... *xi*

Chapter 1   Symptoms and Diagnosis 1

Chapter 2   Causes and Contributing Factors 17

Chapter 3   Treatment and Therapy 39

Chapter 4   Future Prospects 45

Conclusion *55*

Bibliography *57*

Glossary *59*

Index *63*

Acknowledgments

I am grateful to my friend A. Malcolm Campbell for encouraging me to write this book. I appreciate how he advised me to take a leap of faith on this project and on several others that have shaped my career as a science educator. I value Malcom as a teaching and research collaborator, and I am proud of the positive impact that we have made together on science education and the improvement of science literacy.

This book would not have been possible without the support of my wife, Patty Eckdahl. She understands my passion for science and science education and helps me to channel it in ways that benefit students and others around me. I also appreciate the support and encouragement that my parents, Tom and Bonnie Eckdahl, gave me in the pursuit of an education what would give me the privilege of sharing my love of DNA and genetics with undergraduate students and everyone else I meet.

I am grateful to my undergraduate genetics professor at the University of Minnesota, Duluth, Stephen Hedman, for helping to understand that I could pursue my love of genetics in graduate school. Thanks to John Anderson at Purdue University, who taught me to conduct molecular genetics research and to value undergraduate education. I appreciate the environment that Missouri Western State University provided me for following a path to becoming a science educator. I am grateful to my mentors in the Missouri Western Biology Department, Rich Crumley, Bill Andresen, John Rushin, and Dave Ashley, who helped me to learn how to engage students in the classroom and in the research lab. I appreciate the many students that I have had the privilege of teaching in class and collaborating on research projects outside of class. I take pride in the contributions that my former students have already made and will continue to make to society.

Introduction

Four hundred years ago, Europeans would warn, "The child will die whose brow tastes salty when kissed." The ominous saying was not just a part of folklore. A medical research paper published in 1606 described that the fingers tasted salty after touching the forehead of a "bewitched" child. Other references from medical literature of the time connected the salty brows of children to a pancreatic birth defect that prevented them from living past childhood. The salty kiss connects the seemingly innocuous production of excessively salty sweat and a deadly genetic disease. The disease was named "cystic fibrosis of the pancreas" in 1938 by American pathologist Dorothy Anderson, based on her observation of abnormal cysts and fibrous tissues in the pancreases of children who had died of malnutrition. Anderson also described pathology associated with intestines and lungs of these children. Other medical researchers referred to the disease as mucoviscidosis, because of the viscous mucus produced by an afflicted child. They noticed that thickened and sticky mucus clogged the ducts that led from the pancreas to the intestines and created a pasty environment in the lungs that was highly susceptible to infection. An explanation for the salty sweat came in 1948 when Paul di Sant'Agnese, the founder of the Cystic Fibrosis Foundation, discovered abnormally high levels of salt in the sweat of babies with cystic fibrosis (CF). CF is a genetic disease that causes a variety of symptoms associated with the lungs, pancreas, intestines, and liver because of a salt imbalance.

This book presents CF as an inherited disease of the secretory glands that produce mucus and sweat. Chapter 1 describes the symptoms associated with CF that result from the accumulation of abnormally thick mucus in the lungs, including severe breathing problems and bacterial infections. It describes complications of CF associated with lung disease, pancreatic insufficiency, liver disease, intestinal obstruction, and infertility. Chapter 1 explains how a CF diagnosis is made with newborn genetic screening, the sweat test for chloride levels, and measurements of lung function. Chapter 2 describes how CF is caused by six classes of mutations in the *CFTR* gene that encodes the CFTR protein, a gated ion channel that maintains the balance of salt and water in the lungs and other organs. It presents an explanation of the pattern of inheritance of CF along with scenarios in which it can arise in families. Chapter 2 also discusses genetic and environmental factors that affect the progression of CF disease. Chapter 3 describes CF treatments and therapies including antibiotics for infections, medicines that address digestive problems, and respiratory and pancreatic enzyme replacement therapies. It details improved methods for diagnosis of CF such as genetic analysis. It recounts the recent discovery of pharmaceuticals that target the impaired production of CFTR protein and those that are able to rescue its dysfunction. Chapter 4 evaluates the prospects for curing CF through gene therapy and explains how genome editing might be used in the future to correct the *CFTR* gene mutations underlying CF.

CHAPTER 1

Symptoms and Diagnosis

Cystic fibrosis (CF) stems from a change in function of secretory glands, which most people take for granted. Secretory glands include sweat glands, salivary glands, mammary glands, and glands of the digestive system. Mucous glands produce and secrete **mucus,** a mixture of antiseptic enzymes, antibodies, salts, and glycoproteins called mucins. Mucus protects and lubricates layers of cells called the **epithelium,** which lines blood vessels, surfaces, and hollow organs throughout the body. The process by which secretory glands produce mucus is impaired by CF. The normal consistency of mucus is thin and slippery, but CF causes it to be abnormally viscous and sticky. The thick mucus clogs tubes and ducts in the lungs, pancreas, sinuses, liver, gallbladder, intestines, and sex organs. The major symptoms of CF are associated with the respiratory and digestive systems. Respiratory symptoms include inflamed nasal passages, chronic sinusitis, wheezing and breathlessness, intolerance to exercise, a persistent cough that produces thick mucus, and chronic lung infections and bronchitis. Digestive symptoms include abnormal stool consistency that is either too loose and watery or too dry and hard, intestinal blockage, constipation, pancreatic insufficiency, and poor weight gain. Table 1.1 lists the major symptoms of CF and their typical age of onset. Table 1.1 presents a deceptively clear picture of CF, however, since the occurrence, severity, and age of onset of most of the symptoms listed varies widely among people with the disease.

What Can We Learn About CF from Sweat Glands?

The CF dysfunction that causes secretory glands to produce abnormal mucus also affects the function of sweat glands that cover much of the human body. Sweat glands are found in the highest density on the palms,

Table 1.1 *Major symptoms of cystic fibrosis and their typical age of onset*

Typical age of onset	Cystic fibrosis symptom
Prenatal	Hyperechogenic bowel
Newborn	Meconium ileus
	Salty sweat
	Jaundice
Infancy	Recurrent respiratory infections
	Failure to thrive
Childhood	Chronic lung infections
	Chronic sinusitis
	Poor weight gain
	Rectal prolapse
Adulthood	Chronic lung infections
	Pancreatic insufficiency
	Pancreatitis
	Malnutrition
	Chronic bronchitis

soles, and head, where there are as many as 400 glands per square centimeter of skin surface. Sweat glands secrete sweat onto the skin surface through ducts. Sweat is a clear, odorless solution of sodium chloride dissolved in water that helps to control body temperature. Evaporation of the water in sweat carries heat from the skin, thereby cooling it. As sweat moves along the sweat gland duct, most of the sodium chloride is reabsorbed. This results in the production of dilute sweat so that the body can be cooled by evaporation without losing a lot of salt. Sodium chloride reabsorption is facilitated by the epithelial sodium ion channel (ENaC), a protein located on the surface of epithelial cells throughout the body, including the sweat glands, kidneys, lungs, and colon. During sodium chloride reabsorption in sweat glands, sodium ions flow through the ENaC. Chloride ions, the other half of sodium chloride, are transported across the outer membranes of ductal sweat gland cells by flowing through an ion channel formed by a protein called the **cystic fibrosis transmembrane conductance regulator** (CFTR).

People with CF produce normal amounts of sweat, but the salt level in the sweat is abnormally high. This is the basis of the salty kiss.

The elevated salt level comes from the failure of salt reabsorption in the sweat gland ducts. The connection between salty sweat and CF has been known for over four hundred years. The sweat gland may seem to serve an innocuous bodily function, but the study of its dysfunction has led to diagnosing CF and understanding its devastating effects on organs and tissues throughout the body. Sweat gland duct cells have impaired flow of chloride ions through the CFTR ion channel in their cell membranes and lack other ion channels that could rescue the loss of CFTR function. Because the flow of positively charged sodium ions follows the flow of negatively charged chloride ions, impaired flow of chloride ions equates to impaired flow of sodium chloride. The concentration of sodium chloride in sweat is normally 20 to 30 mM compared to 100 mM in CF patients. This large concentration difference makes salty sweat, which is the most reliable single symptom for CF. People with CF lose an unusually high level of salt through sweating, which is exacerbated by heavy sweating during strenuous exercise, fever, or hot weather. Low internal salt levels in people with CF can lead to a variety of symptoms such as fatigue and weakness. CF patients are especially susceptible to heatstroke, which can occur if the body temperature exceeds 104°F. Heatstroke can result in a high fever or unconsciousness and can be an emergency because of the risk of heat-induced brain damage. Low levels of internal salt can cause muscle cramps because of the fundamental role that sodium ions play in muscle cell contraction. People with CF suffer from stomach pain, vomiting, and dehydration due to excessive loss of salt. They also generate a fever more easily.

CF Affects the Respiratory System

The impaired function of the CFTR ion channel to properly transport chloride ions in the sweat glands of CF patients causes serious symptoms, but the problems it causes in the respiratory system can be life-threatening. Our sinuses filter and moisten the air we inhale by producing mucus to trap dust, smoke, or pollen. Mucus also captures potentially pathogenic bacteria and viruses, preventing them from passing into our lungs. Once captured by mucus, the particles and pathogens are moved to the nose by millions of tiny hairs called cilia in a process called **mucociliary clearance**. This process is impaired by CF. The sinus mucus membranes of many CF patients are anatomically abnormal, as shown by increased opaqueness in X-ray images. Most people with CF develop sinus problems between 5 and 14 years of age. Abnormally thick and sticky mucus clogs the sinuses and other passageways that carry air through the upper respiratory system into and out of the lungs. Poor mucociliary clearance causes inflamed nasal passages or a stuffy nose. Thick mucus and damaged cilia in the upper respiratory system often lead to chronic infection of the sinuses, or **sinusitis**. The most common infectious bacteria found in the upper respiratory system of people with CF are *Haemophilus influenzae, Pseudomonas aeruginosa, Staphylococcus aureus, Burkholderia cepacia, Stenotrophomonas maltophilia*, and *Acinetobacter xylosoxidans*. People with CF also carry more harmless bacteria in their sinuses.

The respiratory system includes the sinuses, trachea, and bronchial tubes, all of which are lined with submucosal glands. Submucosal means that the gland resides in the connecting tissue below the mucosa, a skinlike lining of the tubes. Submucosal glands secrete mucus that captures dust, smoke, or pollen particles as well as bacterial and viral pathogens. Cilia move mucus up the bronchial tree toward the mouth to keep the lungs clean and clear. Once mucus accumulates, it is coughed out of the lungs. The CFTR ion channel opens and closes to transport chloride ions out of epithelial cells and into mucus. Positively charged sodium ions follow the chloride ions, increasing the salt concentration of the mucus. Salty mucus makes it hypertonic compared to the inside of the cell, which means that it has more molecules dissolved in it. Hypertonic mucus causes water to move by osmosis out of the cells and into the mucus. Salt regulation has a significant impact on the amount of water in mucus associated with the lungs. The proper regulation of the CFTR chloride channel allows cells to maintain the correct balance of salts on both sides of their outer membranes. When chloride ions cannot leave the cell properly through CFTR, water is retained in the cell because of osmosis and mucus remains thicker than it should be. This causes major problems for CF patients.

Thick and sticky mucus caused by CFTR dysfunction clogs bronchial tubes in the lungs. Mucociliary clearance is impaired by the thick and sticky mucus, causing a chronic cough. The accumulation of thick mucus in the lungs also produces a very favorable environment for infectious microorganisms to live and multiply. CF often leads to pneumonia, a bacterial or viral infection of one of the lungs, or double pneumonia, an infection of both lungs. The immune response to bacterial lung infection releases fluid into the tiny air sacs in the lungs, called **alveoli.** This impairs their gas exchange function and reduces the critical flow of oxygen to organs and carbon dioxide away from them. Infected lungs leak fluids and shed dead cells that clog alveoli and make it difficult for lungs to function. Repeated bacterial lung infections cause **bronchitis**, an inflammation and swelling of the larger airways leading to the lungs, called bronchi. Bronchitis obstructs the airways, making breathing difficult. It also generates excess mucus, providing the opportunity for even more bacterial growth. Persistent bacterial infection in the lungs usually results in a chronic cough that produces thick and colored mucus that is often pus-filled. Thick and dark yellow phlegm accumulates when the immune system sends white blood cells, called **neutrophils**, to the site of the infection. Neutrophils produce an antimicrobial protein called myeloperoxidase, a green protein that can alter the color of phlegm.

The most common bacteria infecting the lungs of people with CF are *S. aureus, H. influenzae, P. aeruginosa, B. cepacia,* and several species of Mycobacteria. Usually, *S. aureus* is the first pathogen to colonize the lungs of children with CF. *S. aureus* pneumonia leads to inflammation of the alveoli, causing a persistent cough and shortness of breath. Other symptoms include fever, chills, and fatigue. Before the use of antibiotics, *S. aureus* was the leading cause of death in children with CF. *H. influenzae* also commonly infects children with CF. *H. influenzae* can undergo hypermutation, which enables it to resist antibiotic treatment even with modern medicines.

The lung colonization of infants and children with *S. aureus* and *H. influenzae* damages epithelial surfaces. This damage can lead to the attachment of *P. aeruginosa* bacteria to the surfaces. The *P. aeruginosa* bacteria may supplant *S. aureus* and *H. influenzae* bacteria. *P. aeruginosa* is a part of the normal microbial population of the respiratory tract, but it can be a pathogen in people with CF. *P. aeruginosa* is a prevalent early pathogen, and the lungs of most children with CF are colonized with it within their first year. Among people with CF above 10 years of age, *P. aeruginosa* is the most common cause of pneumonia. Chronic infection with *P. aeruginosa* is the main cause of respiratory decline in CF patients.

Infection damages the lining of the airways, which can impede the flow of air. The **inflammatory response** of the immune system to bacterial infection is to send neutrophils to the site of infection to combat the bacterial infection with the release of oxidants and enzymes. The immune response further damages the airway. *P. aeruginosa* has the ability to develop **biofilms**, which is key to its ability to survive for very long periods in the lungs. During biofilm formation, bacteria engage in cell-to-cell communication to determine if the requisite density of bacteria exists to form a biofilm. The bacteria produce an extracellular polymeric substance that causes them to stick to each other and to organ surfaces. Other examples of biofilms by different species include dental plaque and the slime on river stones. The development of a pneumonia biofilm may enable the bacteria to become increasingly resistant to antibiotics and immune system host defense. Bacteria in a biofilm can survive 1,000-fold higher antibiotic concentrations than the same bacteria growing as individual cells. Antibiotic-resistant, biofilm-forming *P. aeruginosa* bacteria are thought to play a major role in the progression of lung disease in people with CF.

Nontuberculous mycobacteria (NTM) can also cause lung infections in people with CF. The nontuberculous modifier distinguishes these bacteria from *Mycobacterium tuberculosis*, the causative agent of tuberculosis. Another dangerous relative, *M. leprae*, causes leprosy. Although the societal effects of pulmonary infection by *M. tuberculosis* and *M. leprae* are historically significant, NTM pulmonary infections cause more cases of disease in the modern world. The use of improved culturing techniques and DNA-based identification methods has led to the identification of 150 species of NTM bacteria. Everyone comes into frequent contact with NTM bacterial populations found in drinking water, household plumbing, and drainage water. For this reason, NTM bacteria are also called environmental mycobacteria. NTM bacteria have been found in hospital and dental office water systems. They are also common in peat-rich soils, swamps, and brackish marshes. Since the 1990s, NTM bacteria have been found in the sputum of an increasing fraction of people with CF. The effects of NTM pulmonary infection can vary from no symptoms to severe cough, fatigue, and weight loss. The ability of NTM bacteria to form biofilms gives them resistance to disinfectants and antibiotics, making treatment difficult.

Another group of bacteria that can cause pulmonary infections in CF patients is the *B. cepacia* complex (BCC). There are 17 different BCC species. BCC infections are difficult to recognize because of their similarity to other lung infections. Common symptoms include fever, cough, congestion, wheezing, and shortness of breath. The fact that these symptoms are usually already present in people with CF makes diagnosing BCC very challenging. The severity of the symptoms caused by BCC infection varies widely among CF patients. In many people with CF, BCC infection does not appear to affect their lung symptoms. In others, the rate of lung function decline is only slightly faster. BCC infection can lead to cepacia syndrome, a rapidly fatal pneumonia and bacteremia, or infection of the bloodstream. BCC bacteria spread either by direct contact, such as kissing, or indirectly from touching objects such as doorknobs. Treatment of infection is difficult because BCC bacteria are often resistant to multiple antibiotics. However, some BCC species may be successfully treated with combinations of several antibiotics.

CF and the Digestive System

The production of abnormally thick and sticky mucus caused by CF also impairs essential functions of the digestive system, leading to additional symptoms of the disease. Several of these symptoms are directly related to impaired function of the pancreas. Normally, pancreatic gland cells secrete digestive enzymes that mix with mucus secreted by pancreatic duct cells. The inability of the CFTR ion channel to properly regulate the level of mucus salts causes pancreatic secretions to become thick and sticky, leading to blockage of the ducts that deliver digestive enzymes to the small intestine. The result is **pancreatic insufficiency**, and it occurs in 85 to 90 percent of CF patients. Poor absorption of nutrients from proteins, fats, and carbohydrates caused by pancreatic insufficiency results in malnutrition. The symptoms of malnutrition are poor weight gain and growth. For newborns, malnutrition can result in a condition known as **failure to thrive**, which means that they do not grow at the normal rate. Abnormal digestion of food in the digestive tract leads to loose or "greasy" stool and, sometimes persistent diarrhea. Alternatively, pancreatic insufficiency can cause bulky stools that lead to severe constipation. The frequent necessity for extreme straining while passing stool can cause part of the rectum to protrude outside the anus. This condition is called **rectal prolapse** and may require surgery to repair the damage.

Hyperechogenic bowel is observed in some fetuses with CF. During a second trimester ultrasound examination, the bowel appears unusually bright compared to the surrounding fetal tissues and bones. The ultrasound brightness is caused by abnormal accumulation of meconium in the bowel. Meconium is a thick, greenish black material formed in fetuses from the digestion of swallowed amniotic fluid. Because 10 percent of fetuses with hyperechogenic bowel will have CF, hyperechogenic bowel can contribute to a CF diagnosis. The accumulation of meconium often leads to a CF symptom in newborns, called **meconium ileus**. Meconium ileus is a blockage of the ileum of the large intestine by meconium that occurs in about 18 percent of babies born with CF. Abnormally thick and sticky meconium can make the first bowel movement of a newborn difficult or impossible. Meconium ileus is a life-threatening symptom that must be resolved by medication to break up the blockage, or surgery to remove it.

How Is a CF Diagnosis Made?

Diagnosis of CF is usually made at an early age. Early diagnosis allows families to work with health care providers to treat and manage the disease. Early treatment leads to fewer hospital stays and longer life. Diagnosis of CF usually begins with **newborn screening** in the first 2 or 3 days after birth. Newborn screening for CF occurs throughout the United States, although the method by which it is done varies by state. Some states only require a blood test that measures **immunoreactive trypsinogen** (IRT) produced by the pancreas. Trypsinogen is involved in protein digestion. The effect of CF on the pancreas often causes an elevated level of IRT. The blood test is not definitive, however, since IRT levels can be elevated by premature delivery or stress during delivery. Some states require a CF DNA test in addition to the IRT test. The DNA test checks for mutations in the *CFTR* gene that cause CF. Diagnosis of CF can also be made by simple observation of symptoms. Although each of the symptoms of CF can be caused by other medical conditions, certain combinations of symptoms indicate CF. A characteristic symptom of CF is salty sweat. The elevated levels of salt in the sweat of newborns with CF can be tasted as a salty kiss. CF babies can have persistent diarrhea. They may have difficulty breathing, making a wheezing or crackling sound, or may develop a chronic cough. They may have trouble absorbing nutrients from their food, resulting in failure to thrive. In cases where these symptoms are not present or are not noticed in babies with CF, other symptoms that lead to a CF diagnosis might develop later in childhood. Children may develop rectal prolapse from continued difficulty in passing hard stools. They may have nasal polyps in the lining of the nose that cause them to have difficulty breathing during sleep. They may have clubbing, which is a rounding and flattening of fingers.

Positive results from newborn screening for CF or observation of CF symptoms in newborns or children often lead to the order to perform a **sweat test**. The sweat test is widely regarded as the gold standard for diagnosing CF. The sweat test is painless and can be administered to people of any age, although sometimes it is difficult to get babies to produce enough sweat. The normal chloride concentration of sweat in infants is less than 29 nM. A chloride concentration higher than 60 nM is considered to be diagnostic of CF. An intermediate concentration between 29 nM and 60 nM indicates the possibility of CF. For anyone older than 6 months, the chloride concentration in sweat is normally less than 39 nM. The chloride concentration considered to be diagnostic of CF is 60 nM, as it is for infants. Two consecutive positive sweat test results are considered to be 98 percent sensitive for the diagnosis of CF.

Positive sweat tests for CF usually lead to genetic testing, if it was not already conducted as part of newborn screening. Genetic testing involves the detection of mutations in the *CFTR* gene that are known to cause CF. Although more than 2,000 such mutations have been described, guidelines from the American College of Medical Genetics and the American College of Obstetricians and Gynecologists set in 2012 call for testing of only the 23 most common mutations. If a mutation is discovered in each of the two *CFTR* alleles in a patient, then the patient has CF. However, the connections between *CFTR* mutations and the severity of CF are not entirely clear, because people with the same mutations may have different symptoms and ages of disease onset. If none of the 23 most common mutations are detected, a more comprehensive genetic test may be performed to detect rare *CFTR* mutations.

Diagnosis of CF based on newborn screening, a sweat test, and a genetic test can be further confirmed by other tests. A chest X-ray can show inflammation or scarring of the lungs caused by CF. A sinus X-ray or CT scan can reveal sinusitis. In patients who are at least 6 years old, lung function can be measured with spirometry, which reveals how much air a person can breathe in and out and how fast they can exhale. Forced vital capacity (FVC) is the volume of air that can be exhaled after taking a full breath. The amount of air that can be forcibly exhaled in one second is called FEV_1. The ratio of FEV_1 to FEV, called FEV_1 percent, is commonly used as a measure of lung function. Because CF restricts the flow of air in the lungs, it causes a lower FEV_1 percent value. A measured FEV_1 percent is compared to the value expected for a person of the same height and age. FEV_1 percent values in 8-year-old girls with CF are usually 75 percent of the value expected for girls without the disease. The FEV_1 percent value falls to 63 percent by 20 years of age. Eight-year-old boys typically have FEV_1 percent values that are 90 percent of the value expected for boys without the disease. The FEV_1 percent value falls to 60 percent by 20 years of age. In addition to its use as a diagnostic tool, FEV_1 percent is used to follow the course of CF disease in patients, and to determine the extent to which they respond to treatments and therapy. More complicated lung function tests may also be done. For example, the total lung capacity (TLC) and residual volume (RV) may be used to determine the volume of air that a person cannot exhale. RV increases when there is damage to the lungs caused by CF.

Another diagnostic laboratory test for CF is a sputum culture for detecting bacterial infections. The patient is asked to rinse his or her mouth and gargle with sterile water to clear most of the bacteria from the oral cavity and the throat before coughing up some sputum into a sterile collection container. The sputum sample is taken to a clinical microbiology laboratory to see if it contains one or more of the bacteria that commonly cause respiratory infections in CF, such as *P. aeruginosa*, *S. aureus*, *H. influenzae*, and *B. cepacia*. Diagnosis of CF can also be confirmed with various blood tests. A full blood count measures the number of red and white blood cells and platelets in the blood. Increased levels of white blood cells may indicate infections that could be caused by CF. Liver function blood tests can detect impaired liver function caused by CF. Blood test for urea and electrolytes, fasting glucose, and vitamins A, D and E may also be performed.

Complications from CF

Because CF affects the function of cells throughout the body, it results in a variety of complications for people living with it. Table 1.2 lists the organs and systems in which CF complications usually occur.

CF lung complications are particularly important because lung disease causes about 80 percent of deaths among CF patients. **Chronic obstructive pulmonary disease** (COPD) occurs when the bronchi and **bronchioles** are narrowed because of inflammation caused by chronic bacterial or fungal infections. COPD is a progressive disease that causes

Table 1.2 Complications from cystic fibrosis

Affected system or organ	Cystic fibrosis complication
Lungs	Chronic obstructive pulmonary disease (COPD)
	Bronchiectasis
	Hemoptysis
	Pneumothorax
	Cor pulmonale
Upper respiratory	Chronic sinusitis
	Nasal polyps
Pancreas	Chronic pancreatitis
	Cystic fibrosis–related diabetes mellitus (CFRD)
Gallbladder	Gallstones
	Small gallbladder
Liver	Cystic fibrosis–associated liver disease (CFLD)
Small intestine	Distal intestinal obstruction syndrome (DIOS)
Large intestine	Intussusception
Reproductive system	Male infertility from congenital bilateral absence of the vas deferens (CBAVD)
	Female difficulty in getting pregnant

coughing, wheezing, and difficulty in exhaling, which results in short-ness of breath. The term COPD includes both emphysema and chronic bronchitis. Emphysema is caused by damage to the walls that separate alveoli within the lungs. The alveoli lose their shape and cannot hold normal volumes of air. Alveoli may fuse to produce larger air sacs with lower surface area-to-volume ratios, which reduce the efficiency of oxy-gen and carbon dioxide exchange. Bronchitis is inflammation of the lin-ing of the respiratory system that causes the lining to become thickened. Bronchitis leads to the production and accumulation of thick mucus that obstructs the airway. COPD from CF can be treated but cannot currently be cured. Bronchodilators can be used to relax the smooth muscles in the airway, and steroid therapy reduces inflammation associ-ated with COPD.

Respiratory complications of CF can lead to acute respiratory failure. Acute respiratory failure occurs when accumulated fluid in the alveoli prevents them from properly exchanging oxygen and carbon dioxide. Symptoms include extreme breathing difficulty, a bluish coloration of the skin, elevated heart rate, profuse sweating, and confusion. Acute respira-tory failure may be treated with supplemental oxygen therapy, broncho-dilators, antibiotics, and airway clearance measures. If acute respiratory failure is life-threatening, lung transplantation is considered. However, not all patients are good candidates for lung transplants.

Lung damage caused by CF can result in protracted abnormally low level of oxygen in the blood. The right ventricle of the heart and the pul-monary artery that connects it to the lungs compensate for low oxygen with an increase in blood pressure. Blood pressure is also increased in the blood vessels inside the lungs, a condition called pulmonary hyper-tension. The resulting strain on the right side of the heart causes **cor pulmonale**, an enlargement of the right side of the heart that can eventu-ally lead to complete heart failure. Symptoms include chest pain, fainting, swelling of the extremities, and a bluish color of the skin. Cor pulmonale is treated with pharmaceuticals and oxygen therapy. A lung transplant or a heart–lung transplant might be the only cure for this complication.

Respiratory problems caused by CF eventually can lead to permanent damage to the bronchi, referred to as **bronchiectasis**. The combination of chronic coughing and repeated infections with associated inflammatory immune responses causes the bronchi to dilate and lose their ability to clear mucus. Enlarged airways become filled with mucus, which makes them susceptible to even more infection, exacerbating the problem. Lung damage caused by bronchiectasis is permanent. People with CF can also suffer from a complication called **hemoptysis,** or coughing up blood. Chronic lung infection requires constant attention from the immune system. One result of this is angiogenesis, or the formation of new blood vessels. The associated delivery of a large volume of blood can cause the blood vessels in the bronchioles to swell and burst. Hemoptysis can lead to airway obstruction, shock, asphyxiation, and exsanguination, the severe loss of blood. In severe cases, it may be necessary to carry out bronchial arterial embolization to stop the bleeding. In this procedure, bleeding is curbed by deliberately creating an embolism, or blockage of a blood vessel.

A life-threatening CF complication called a **pneumothorax** occurs in 10 percent of CF patients. A pneumothorax occurs when air accumulates outside of the lung, causing it to become disconnected from the wall of the chest cavity. Air outside of the lungs puts pressure on the lungs and causes all or part of them to collapse. If only a part of a lung is collapsed, the extra air can be reabsorbed in a couple of weeks without any intervention apart from supplemental oxygen therapy. If a larger part of the lung collapses, a needle or chest tube can be used to remove the air from the pleural cavity.

Another CF complication is allergic bronchopulmonary aspergillosis (ABPA). Nearly half of all CF patients have a fungus in their sputum known as *Aspergillus fumigatus,* and about 10 percent of these people will develop ABPA. The fungus elicits a response by the immune system called hypersensitivity, which leads to inflammation and dilation of the airways. ABPA causes coughing, shortness of breath, and exercise intolerance. The immune response can also lead to a pneumothorax.

Complications of CF can cause problems in the upper respiratory system as well. In people with CF, the sinus mucus membranes that moisten and cleanse inhaled air can develop **nasal polyps.** Nasal polyps are soft benign growths in the nasal passages or sinuses that can affect breathing, cause nasal congestion, impair the sense of smell, and cause a higher risk of bacterial infection. Sometimes nasal polyps protrude from the nostrils.

The frequency of occurrence of upper respiratory polyps in people with CF varies, but it increases with age. Nasal polyposis occurs in about 15 to 20 percent of all CF patients, compared to 4 percent of the general population. Although some polyps can be treated with steroids, many require surgical removal. The polyps may recur afterward. Thick mucus and damaged cilia in the upper respiratory system can lead to chronic infection of the sinuses, or sinusitis. Sinusitis causes chronic congestion, runny nose, nasal obstruction, and postnasal drip. This leads to a constant need to clear the throat and coughing that is aggravated by lying down, especially after sleeping. Sinusitis also causes headaches that do not respond well to medication. Chronic sinusitis early in life is thought to impede the normal development of the frontal sinuses. As a result, adolescents with CF often have no frontal sinus cavity at all.

CF can also lead to complications associated with the pancreas. Blockage of the pancreatic duct into the small intestine can cause scarring and inflammation that results in **pancreatitis**. Chronic pancreatitis inhibits the production of enzymes needed for proper digestion. As a result, the small intestine is unable to properly digest food, which leads to weight loss, abdominal pain, diarrhea, nausea, and abnormal stools. Treatments of chronic pancreatitis include prescribing pain medication, administering intravenous fluids, taking supplemental digestive enzymes, and adopting a specialized diet. Chronic pancreatitis also decreases the ability of the pancreas to produce and secrete insulin into the bloodstream. The resulting complication is **cystic fibrosis-related diabetes mellitus** (CFRD). About 20 percent of CF patients develop diabetes by the age of 30 years. Because CFRD is characterized by the failure of the pancreas to produce normal levels of insulin, it is similar to type 1 diabetes. However, because patients with CFRD also do not respond normally to insulin, CFRD is also like type 2 diabetes. The abnormally high blood glucose levels caused by CFRD result in the symptoms of increased thirst and urination. CFRD also causes fatigue and weight loss. The treatment for CFRD is careful control of blood glucose levels through diet, exercise, and insulin therapy.

CF also can result in abnormal control of sodium and chloride ion transfer across the bile duct from the liver and gallbladder to the small intestine. The bile becomes abnormally thick and concentrated and can block the bile duct. Blockage can result in gallstones and an unusually small gallbladder. Over long periods of time, the blockage can lead to a complication called **cystic fibrosis-associated liver disease** (CFLD). CFLD affects about 30 percent of people with CF. About 3 percent of CF patients die from CFLD, making it the third leading cause of CF death. CFLD results in jaundice, a yellow discoloration of the skin caused by elevated serum levels of bilirubin produced from the breakdown of red blood cells. Other symptoms of CFLD are abdominal pain, inability to gain or maintain weight, and fatigue. CFLD can ultimately result in cirrhosis of the liver, which is damage and scarring of the liver that lead to its failure. Liver cirrhosis is a complication in 1 to 2 percent of all CF patients. CFLD and liver cirrhosis can cause the liver and the spleen to become enlarged and can increase blood pressure in the portal vein leading to the liver, resulting in portal hypertension. Portal hypertension results in swelling of the spleen, accumulation of fluid in the abdominal cavity, and bleeding in the esophagus or stomach.

Another CF complication involving the digestive system is **distal intestinal obstruction syndrome** (DIOS). DIOS is a blockage of the intestines by undigested food and thickened mucus. Because it is caused by the failure of the pancreas to produce and secrete digestive enzymes, DIOS is more prevalent in CF patients who have pancreatic insufficiency. About 5 to 10 percent of all CF patients develop DIOS. The symptoms of DIOS include severe abdominal pain, vomiting, and weight loss. It can result in a mass of accumulated hard feces that can be felt in the abdomen and confirmed by X-rays. DIOS can be treated with medications such as pancreatic enzymes, laxatives, and stool softeners. A common treatment is drinking an electrolyte and polyethylene glycol solution called GoLYTELY, which is also used for bowel cleansing in preparation for a colonoscopy. Surgery may be required to remove the intestinal obstruction.

Intussusception can also be a complication of CF. This occurs when one section of the large intestine slides inside an adjacent section. Intussusception impedes the flow of digested food through the large intestine and can result in blockage. It also damages the nerves and blood vessels associated with the large intestine, resulting in swelling and decreased blood flow. In extreme cases, the lack of blood flow can cause tissue death, resulting in internal bleeding and risk of abdominal infection. Intussusception occurs in about 1 percent of all children with CF. Its symptoms are abdominal pain and vomiting. An intussusception can sometimes be corrected with an enema. Otherwise, surgery may be required to remove the obstruction and any dead tissue and repair the large intestine.

Complications of CF can affect the reproductive system too. About 95 percent of men and 20 percent of women with CF are infertile. Infertility in men with CF is caused by blockage of the vas deferens, the tube that carries sperm from the testicles to the penis. Men with CF can have normal erections and ejaculation, but if the vas deferens is blocked, the semen does not contain sperm cells. In women with CF, infertility is caused by thick mucus blocking the cervix, the opening connecting the vagina and uterus. The blockage can prevent sperm cells from traveling to the fallopian tubes, where fertilization occurs. Assisted reproductive technology (ART) can be used to allow men and women with CF to conceive their own biological children.

CHAPTER 2

Causes and Contributing Factors

How Is Cystic Fibrosis Inherited?

The pattern of inheritance for cystic fibrosis (CF) is simple Mendelian **autosomal recessive**. This pattern was first described in 1866 by Gregor Mendel. Mendel discovered the concept of the gene as a unit of hereditary information. He also discovered that genes can exist in different forms called **alleles** and that one allele can be dominant over another when both are present in the same individual. The groundbreaking insights by Mendel were used in 1946 by Anderson and Hodges to establish the hereditary basis of CF. They considered the alternative hypotheses of dietary deficiency or infection during pregnancy or infancy, but found compelling evidence in favor of a hereditary basis for the disease. Anderson and Hodges completed an analysis of 20 families affected by CF. They found that the ratio of unaffected children to children with CF was about 3:1, and that boys and girls had the disease with equal frequency. This led Anderson and Hodges to hypothesize that CF follows a pattern of simple Mendelian inheritance. They proposed that inheritance of CF is monogenic, involving a single gene. They realized that there must be at least two alleles of the gene and that the disease-causing allele is recessive to the dominant normal allele. The fact that CF affected equal numbers of males and females meant that the gene involved must not be on the X or the Y chromosomes involved in the determination of gender. The 22 human chromosomes that are not the X and the Y chromosome are called autosomes. The genetic basis of CF is captured by saying that it is inherited in a simple Mendelian fashion as a monogenic autosomal recessive disease. For the 20 families studied by Anderson and Hodges, and for hundreds of thousands of families since, this pattern meant that unaffected parents of children born with CF are both **carriers** of the disease-causing allele. A carrier does not have CF because they possess one copy of the normal *CFTR* allele, and complete Mendelian dominance ensures that one copy is sufficient to prevent disease. Figure 2.1 shows a cross between two parents who are both CF carriers. The figure uses a "D" for the dominant normal allele and a "d" for the disease-causing recessive CF allele. It also uses a Punnett square to illustrate the possible offspring genotypes. The term **genotype** refers to the alleles that an individual inherits, and the term **phenotype** describes the trait that results from a genotype. Because there is an equal probability that a sperm cell contains the dominant D allele or the recessive d allele, and because the same is true for egg cells, each of the four possible fertilization events is equally likely. Therefore, the probability of a given offspring having the unaffected genotype of DD is ¼. The probability of the offspring being an unaffected carrier with a genotype of Dd is ½ and the probability of the offspring having the CF genotype of dd is ¼. This applies to each fertilization event, or conception.

The majority of people who are carriers for CF never learn that they are carriers. This is because being a carrier is uncommon, and the probability of two prospective parents both being carriers is even more rare. For many people, however, the frequency of being a CF carrier is higher

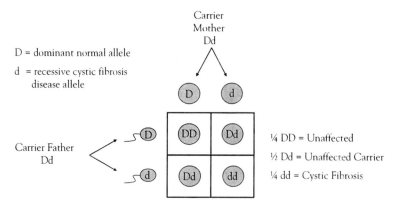

Figure 2.1 Punnett square showing offspring genotypes from two cystic fibrosis carrier parents

than it is for other genetic diseases. For people of European heritage, the probability of being a CF carrier is 1 in 25. Carriers are less common in people of other heritages. About 1 in 46 Hispanics, 1 in 65 Africans, and 1 in 90 Asians is a carrier for CF. Each of these statistics is derived from the incidence of CF in people with a given heritage. For example, the frequency at which CF occurs in the children of parents with European heritage is about 1 in 2,500. This number can be understood to result from multiplying the probability of each of the two parents being a carrier and the probability that the child will inherit the disease-causing allele from each of them. Because the expression (carrier probability $\times$ carrier probability $\times$ ¼) must equal the disease frequency of 1/2,500, the carrier probability is 1/25. Similar calculations reveal that the occurrence of CF is 1 in 8,400 for Hispanic parents, 1 in 17,000 for African parents, and 1 in 32,000 for Asian parents.

The most common way that two parents find out with certainty that they are carriers for CF is that they have a child born with the disease. However, there are other scenarios in which people can learn that their chance of being a carrier are higher than that of the general population. A person could learn that their sibling had a child with CF. Because the sibling must be a carrier for the disease, it follows that at least one parent of the siblings must be a carrier too. Because there is a 50 percent chance that the person inherited the CF allele from their carrier parent, the person has a 50 percent probability of being a carrier. This is considerably higher than that for the general population. If the person is of European descent, the odds are 12.5 times higher, because 50 percent is 12.5 times larger than 1/25, or 4 percent. The odds are 23 times higher for a person of Hispanic descent, 32.5 times higher for Africans, and 45 times higher for Asians. Another common scenario in which a person could discover that they have a higher probability of being a carrier than the general population is that the person has a sibling with CF. In this case, the parents of the two siblings must both be carriers, and so the scenario illustrated in Figure 2.1 applies. Because the person does not have CF, one of the three remaining fertilization events must have taken place when they were conceived. Because two of these three possibilities result in a carrier genotype of Dd, the probability of the person being a carrier is 66.7 percent. This is considerably higher than the carrier probability for any ethnic group.

A less common scenario is that a person learns their cousin had a child with CF. In this case, pedigree analysis can be used to calculate that the person has a 25 percent chance of being a carrier. Again, this probability is higher than the carrier probability for any ethnic group. Even less common is the scenario in which a parent with CF and an unaffected parent have a child with CF. In this case, the fact that the child inherited a disease-causing recessive allele from both parents means that the parent who does not have CF must be a carrier. The probability of each of their children having CF is 50 percent.

When pedigree analysis shows that someone has a higher-than-usual probability of being a CF carrier, they may choose to get a genetic test to find out for sure. Because of the high frequency of carriers in the general population, the American Congress of Obstetricians and Gynecologists recommends that all couples get tested for being CF carriers, even if there is no evidence of CF in the family. Although more than 2,000 *CFTR* mutations known to cause CF have been described, guidelines from the American College of Medical Genetics and the American College of Obstetricians and Gynecologists call for testing for the 23 most common mutations. Because not all the possible mutations are analyzed, testing results cannot eliminate the chance that someone is a CF carrier. The distribution of disease-causing mutations also varies among groups of people of different ethnic backgrounds. Some groups are more likely to have the 23 mutations that are routinely used in carrier testing. As a result, the rate of correct detection of CF carriers with the 23-mutation panel is higher for some groups than others. The detection rate is 88 percent for people of European descent, 72 percent for Hispanics, 64 percent for Africans, and 49 percent for Asians. Nonetheless, a negative test for the 23 mutations does reduce the probability that someone is a CF carrier to 1/170 or lower for people of all ethnic backgrounds. A positive test result for one of the 23 mutations is followed up with confirmation of the result by determination of the DNA sequence of both *CFTR* alleles.

Positive results of CF carrier testing provide couples with information that they can use to make decisions about having a family. A couple that learns that each of them is a carrier for CF may decide not to risk conceiving a child that has CF, or even one that is a CF carrier. They might choose to have a family by adopting a child. The couple might decide to avoid the risk of having a child with CF through the use of donor sperm or eggs. For practical reasons, the choice of using donor sperm is more common. Assuming the donated sperm or egg cells do not themselves carry an allele that causes CF, the risk of having a child with CF is eliminated. Another option for the couple is to use **preimplantation genetic diagnosis** (PGD). This method begins with in vitro fertilization, during which several eggs from the woman are harvested and mixed with sperm from the man. The resulting fertilized eggs are allowed to develop into young embryos. Several cells from each of the embryos are extracted as a tiny biopsy. The DNA from the cells is subjected to genetic testing for *CFTR* mutant alleles. Embryos found to contain two mutant alleles would develop into children with CF and are therefore discarded. Embryos with one mutant allele would develop into CF carriers and are also avoided. Among the remaining embryos, one or more that do not have any mutant alleles are chosen for implantation and the establishment of a pregnancy. Yet another option is for the couple to become pregnant and then have prenatal genetic testing done to determine if the fetus carries two mutant *CFTR* alleles that would cause it to develop into a child with CF. This test requires amniocentesis or chorionic villi sampling to obtain fetal cells that can be analyzed. The test results can be used by the couple to make the decision to abort the fetus. Alternatively, they can use the results to prepare themselves and their health care providers for the birth of a child with CF.

What Is the Molecular Basis of CF?

Assigning the cause of CF to mutations in the *CFTR* gene began with the observation of high levels of salt in the sweat of patients with the disease. This led to the discovery of abnormal transport of chloride ions in the sweat gland cells of people with CF. The basis for the defective chloride ion channels became clearer when the *CFTR* gene was identified using RNA molecules found to be abundant in diseased CF tissues. The identification of the *CFTR* gene as the cause of CF was supported by the observation that mutant alleles of the gene were present in people with CF. The *CFTR* gene sequence was found to encode a protein of 1,480 amino acids. Because the structure of the CFTR protein did not clearly reveal its function, it was called the CF transmembrane conductance regulator, or CFTR. This name was intended to describe its suspected role as a channel or a regulator of channels. It is now known to be both.

The *CFTR* gene is one of about 2,100 genes found on chromosome number 7, which are a subset of the approximately 22,000 protein-coding genes in the **human genome**. The *CFTR* gene spans about 189,000 base pairs of the almost 160 million base pairs of chromosome 7. The *CFTR* gene is large, as the average size of a human gene is about 54,000 base pairs. As is typical of human genes, only a fraction of the *CFTR* gene codes for the CFTR protein itself. The protein-encoding DNA sequences of 98 percent of human genes are split into segments called exons, named for expressed regions. The average number of exons in human genes is about 10 and their average size is 288 base pairs. The *CFTR* gene has more exons than most genes, with 27 exons, but nowhere near the record number of 363 exons in the *titin* gene. Exons are interspersed with introns, and so the *CFTR* gene has 26 introns interspersed between its 27 exons. In order for the information in a gene to be used to make a protein, the DNA of exons and introns alike is first used to produce RNA in a process called **transcription**. RNA produced by transcription undergoes a process called **RNA splicing** that removes all of the introns and splices together successive exons to make a contiguous protein-encoding sequence. For the *CFTR* gene, splicing together the 27 exons results in a messenger RNA (mRNA) that is 6,129 base pairs in length. The mRNA is transported from the nucleus, where transcription and RNA processing occur, to the part of the cell outside the nucleus called the cytoplasm, where translation occurs.

Translation is the process by which the base sequence of mRNA transcripts is used to direct the synthesis of proteins. Amino acids are the building blocks of proteins. There are 20 naturally occurring amino acids and their sequence in a protein determines its structure and function, or dysfunction. The bases of an mRNA are translated in groups of three, called codons. Because there are four possible bases in each of the three positions of a codon, there are $4 \times 4 \times 4$, or 64 different possible codons. Groundbreaking genetics, biochemistry, and molecular biology experiments in the 1960s revealed which amino acid in encoded by each of the 64 codons. This information is referred to as the genetic code.

One of the codons is a start codon for the initiation of translation, and three are stop codons that signal the end of translation. Translation occurs in two places in the cytoplasm of cells, depending on the destination of the protein produced. For proteins that function inside the cell, such as enzymes or the proteins that make up the cytoskeleton, translation occurs on free ribosomes floating in the cytoplasm. For proteins that are either embedded in membranes or that localize inside membrane-bound cellular compartments such as secretory vesicles or lysosomes, translation occurs on ribosomes that are associated with a system of membranes called the endoplasmic reticulum (ER). Folding of proteins is influenced by the addition of sugar groups in a process called cotranslational modification and by interaction with specialized folding proteins referred to as chaperones.

Translation of the *CTFR* mRNA occurs in association with the ER because the CFTR protein is embedded in cellular membranes. The 6,129 base pair CFTR mRNA is used to produce a chain of 1,480 amino acids that folds into the CFTR protein, which contains the seven domains illustrated in Figure 2.2. The cytoplasmic amino terminus is the first part of the protein to be made. The term "amino" in its name refers to the fact that polypeptide chains have a directionality. The amino acids are linked in an orientation that produces an amino end and a carboxylic acid end. The term "cytoplasmic" indicates that the cytoplasmic amino terminus is in the cytoplasm, or inside of the cell. The cytoplasmic carboxyl terminus

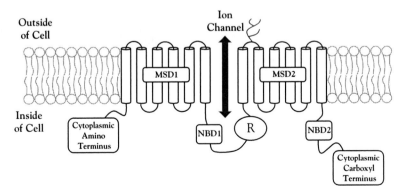

Figure 2.2 Structure of the CFTR protein showing its structural domains and ion channel

is the last part of the protein to be made. The cytoplasmic amino terminus and cytoplasmic carboxyl terminus contain specific amino acids that enable them to bind to, and interact with, other proteins inside the cell. The CFTR has two membrane-spanning domains (MSDs) composed of a series of six alpha helices that insert into the membrane. Alpha helices are fundamental units of protein structure in which a chain of amino acids forms a right-handed coil. The CFTR MSDs form the channel through which negatively charged ions such as chloride can flow into and out of cells. The MSDs also contain a gate that controls the flow of ions. The CFTR protein undergoes cotranslational modification that adds branched sugar chains to one of the MSD extracellular loops. The sugar chains are involved in CFTR protein folding, stability, and trafficking to the outer cell membrane. Shortly after each of the two MSDs are two domains that interact with adenosine triphosphate (ATP). ATP is a nucleotide that used throughout the cell to provide energy for its metabolic processes. The two domains are called the nucleotide-binding domains (NBD1 and NBD2). The regulatory domain, or R domain, is found on the cytoplasmic side of the membrane between the two NBDs. CFTR function is controlled by cycles of ATP binding and hydrolysis at the two NBDs and phosphorylation of the R domain. The CFTR uses the energy of ATP hydrolysis to adopt either an ion-conducting or a nonconducting shape. The R domain can be modified with phosphorylation by enzymes called kinases that transfer a phosphate group from ATP bound to the NBDs onto amino acids in the R domain. Phosphorylation of CFTR opens the ion channel gate. Dephosphorylation by enzymes called phosphatases closes the ion channel gate.

The CFTR protein acts as a chloride ion channel of low conductance capacity with no preference for the direction of ion transport. This enables the CFTR to play roles in both absorption and secretion of ions. Once CFTR is produced by translation in association with the ER, the protein moves to the outer cell membrane of secretory and absorptive epithelial cells of the sweat glands, airways, pancreas, liver, gastrointestinal tract, and vas deferens. The CFTR ion channel allows the transport of chloride ions from one side of the outer cell membrane to the other. The movement of chloride ions in these epithelial cells changes the water concentration of the fluids in ducts and organs. Osmosis, or the flow of water across membranes, occurs because of the differences in ion concentrations across outer cell membranes. Osmosis results in fluid movement and secretion of fluids throughout the body. Sweat is produced in sweat glands. Mucus moistens the upper and lower respiratory tract. Pancreatic fluids are released to the small intestine. Fluids are released that contribute to bile secretions. The CFTR protein also regulates other ion channel proteins, plays a role in transporting ATP, controls the movement of substances into and out of cells, and regulates the pH of cellular compartments.

The proportion of CFTR proteins that function as active ion channels in the membrane of epithelial cells is regulated by ATP nucleotide binding and subsequent phosphorylation. The total number of CFTR protein molecules present in the membrane is also regulated by the cell. A first level of regulation occurs in the ER, where CFTR folding undergoes quality control. Abnormally folded mutant CFTR is kept in the ER and degraded by a process known as ER-associated degradation (ERAD). The process involves a cellular protein complex called the proteasome, which has a pore through which proteins pass for degradation by protease enzymes. For CFTR proteins that are properly folded and arrive at the outer cellular membrane, a second level of regulation occurs that controls how many of the CFTR protein molecules stay in the outer cell membrane and how many are retained inside the cell.

CFTR Gene Mutations Cause CF

The causes of CF are mutations in the *CFTR* gene. The mutations disrupt the normal process of gene expression that produces the cell-type appropriate number of functional CFTR proteins in epithelial cells of the sweat glands, airways, pancreas, liver, gastrointestinal tract, and vas deferens. The most common mutation of the *CFTR* gene is a deletion of three bases in exon 11. Normally, the sequence ATCTTT in the DNA of the *CFTR* gene includes an ATC codon that results in the incorporation of an isoleucine amino acid into the polypeptide chain and a TTT codon that dictates incorporation of a phenylalanine amino acid. The mutation is a deletion of the three bases CTT from the ATCTTT sequence in the DNA of the *CFTR* gene. This results in the sequence ATT, which also encodes isoleucine, a result of the redundancy of the genetic code. However, the codon that specifies phenylalanine is missing, and so the resulting polypeptide is missing a phenylalanine. The mutation has a name that reflects the change in the DNA sequence and an alternative name that describes the change in the amino acid sequence of the protein. The DNA mutation name c.1521_1523delCTT begins with a c to refer to the coding sequence of the mRNA produced by transcription of the gene. It specifies the base positions of 1521 to 1523, contains a del for deletion, and identifies the deleted bases as CTT. The protein mutation name is p.Phe508del, which includes a p for protein, a Phe for the phenylalanine amino acid affected, 508 for its position in the polypeptide chain, and del for its deletion. Although the legacy name for the mutation is F508del (where F stands for phenylalanine) and is widely used, this book will refer to the mutation by the name of p.Phe508del. Although the deletion of a single phenylalanine in a chain of 1,480 amino acids seems at first to be a minor change, it has a profound effect on the function of the CFTR protein. The mutant protein does not fold properly and is degraded before it can be transported to the outer cell membrane. Degradation happens as a result of the ERAD system by which misfolded proteins are destroyed. Cells carefully monitor the folding of CFTR proteins, as indicated by the observation that 75 percent of nonmutated CFTR is degraded before it leaves the ER. When cells make mutant CFTR protein p.Phe508del, 100 percent of it is degraded. As a result, the mutant protein does not reach the outer cell membrane and thus cannot carry out its role as an ion channel and a regulator of other ion channels. Lack of CFTR ion channels results in abnormal control of fluids secreted by epithelial cells.

The p.Phe508del mutation is the major contributor to worldwide CF. However, advances in molecular genetics technology have enabled researchers to identify over 2,000 mutations of the *CFTR* gene that have a variety of effects on CFTR protein production and function. Mutations in any gene can have one of the three possible effects. They can be deleterious, with negative impacts on health that range from minor to fatal. Deleterious mutations attract the most attention because of the obvious connection to health. Mutations can also be beneficial, providing an advantage to individuals who carry them. These mutations have accumulated in the human genome and have shaped the repertoire of alleles that constitute our gene pool. The third type of mutation is called neutral and it is the largest category. Neutral mutations have no effect on protein production or function. Of the over 2,000 *CFTR* mutations described so far, 82 percent have known or predicted deleterious effects. For 14 percent of the mutations, no effect has been observed or is predicted, and so they can be described as neutral mutations. For the remaining 4 percent of mutations in the database, the effects on protein production or function are unknown and unpredictable. For mutations that cause an adverse effect on production or function of the CFTR protein, it is useful to group them into six classes. The six classes organize *CFTR* mutations according to the specific part of the pathway of gene expression that they affect. Figure 2.3 illustrates the effect of mutations in the six classes on production of functional CFTR protein. Mutations in categories I, II, and III are associated with severe symptoms of CF, such as pancreatic insufficiency. Class IV and V mutations result in milder symptoms. People who possess two mutant alleles from categories IV and/or V usually have normal pancreas function, mild disease, and lower concentrations of salt in their sweat. People with one class IV or V mutation and one mutation from categories I, II, or III usually have normal pancreas function. This means that the milder mutations are dominant over the more severe ones. Class VI mutations result in decreased CFTR stability and are associated with severe functional and phenotypic consequences.

Class I *CFTR* mutations result in unstable CFTR mRNA and a total or partial absence of the CFTR protein. These mutations introduce or use a stop codon in the mRNA that is referred to as a premature stop codon or a premature termination codon. Because translation of an mRNA containing a premature stop codon would produce a shortened protein, cells have a system to degrade the mRNA. The system is called nonsense-mediated RNA decay. A nonsense mutation is one in which a codon that encodes an amino acid is changed to one of the three stop codons. Because the errant stop codon is inappropriately placed within the coding sequence, the stop codon is referred to as nonsense. Class I mutations are of three types, namely nonsense mutations, frameshift mutations, and splicing mutations. An example of a class I nonsense mutation is c.1624G>T, a change from G to T at base 1,624 of the *CFTR* mRNA that converts a glycine codon to a stop codon. As a result, the mutation

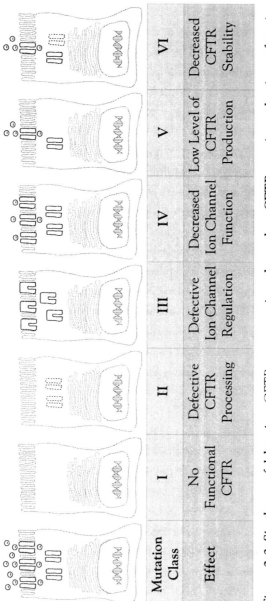

Mutation Class	I	II	III	IV	V	VI
Effect	No Functional CFTR	Defective CFTR Processing	Defective Ion Channel Regulation	Decreased Ion Channel Function	Low Level of CFTR Production	Decreased CFTR Stability

Figure 2.3 Six classes of deleterious CFTR gene mutations that reduce CFTR protein production or function

is also called p.Gly542X, where the X refers to the stop codon. The 542 in the protein name of the mutation indicates that translation would produce a polypeptide chain of only 542 amino acids in length instead of the normal length of 1,480 amino acids. However, the c.1624G>T mRNA is destroyed by the nonsense-mediated RNA decay system before it can be translated.

Class I *CFTR* frameshift mutations occur when deletion or insertion of bases disrupts what is called the translational reading frame, which establishes how mRNA bases are read during translation in groups of three. Insertions or deletions of multiples of three bases do not cause frameshift mutations. An example of a *CFTR* class I frameshift mutation is c.1650delA. The deletion of one A at position 1,650 in the coding sequence of the mRNA results in a translational frameshift. All of the amino acids in the protein are normal up to position 551, after which the amino acid sequence changes. The altered reading frame encounters a stop codon eight codons after the deletion. The mutant protein name p.Gly551ValfsX8 reflects the molecular consequences of the single base deletion. Translated p.Gly551ValfsX8 would be a shortened version of the CFTR protein that has only 551 amino acids instead of the full complement of 1,480 amino acids. In cells, however, translation does not occur because the mRNA is degraded by the nonsense-mediated RNA decay system.

The third type of class I *CFTR* mutations is splicing mutations. These mutations change one or more of the bases needed for normal mRNA splicing prior to translation. An example of a splicing mutation is c.3140-26A>G, which converts the sequence AA to AG. Aberrant splicing involving the new AG results in the insertion of 25 extra bases from the intron into the resulting mRNA. Because 25 is not a multiple of three, this mutation results in a translational frameshift. It also introduces a premature stop codon. As with other class I *CFTR* mutations, the nonsense-mediated RNA decay system destroys the mutant mRNA. Overall, the mechanism of dysfunction associated with class I mutations is degradation of the mRNA. However, there are class I mutations that produce stable mRNAs with premature termination codons that are translated. These mutations produce a shortened version of the CFTR protein that is unstable and is degraded by the cellular proteasome.

The functional outcome of truncated proteins is the same as the functional outcome of mutations that result in mRNA degradation.

Class II *CFTR* mutations result in abnormal protein folding, processing, and trafficking. Class II *CFTR* mutations represent the majority of the mutant *CFTR* alleles described so far. Class II mutations result in an abnormally low level of CFTR protein in the outer cell membrane. The p.Phe508del mutation that constitutes 70 percent of *CFTR* mutations worldwide is the prototype for class II *CFTR* mutations. As described earlier, the mutation results in the deletion of a phenylalanine that results in a misfolded CFTR protein that is degraded by the cellular proteasome. Scientists do not fully understand how a single missing amino acid among the 1,480 amino acids in CFTR results in aberrant folding, because protein folding is a very complicated process. Protein folding converts the primary sequence of amino acids into characteristic patterns called secondary structures, the most common of which are alpha helices and beta sheets. The secondary structures are in turn folded into a tertiary structure that is called the native state. The native state of a given protein depends more heavily on the identity and presence or absence of particular amino acids in the polypeptide chain than others. There is a long-standing and ongoing research effort to learn the complex rules by which the sequence of amino acids dictates the folding of proteins. For the p.Phe508del mutant CFTR protein, the absence of a phenylalanine eliminates one or more interactions that are needed for the formation of the native state of the CFTR protein. Improper folding of a given p.Phe508del CFTR protein molecule usually leads to its degradation. Alternatively, the mutant protein might reach the outer cell membrane, although it does not stay there as long as the normal CFTR protein. The misfolded protein also spends an abnormally small amount of time in the open gate conformation. The genotype that includes two p.Phe508del alleles results in severe CF symptoms of impaired lung function and pancreatic insufficiency. It is also possible for someone to have one p.Phe508del *CFTR* allele and a second allele that is one of the other 2,000 mutant alleles. This genotype is referred to as compound heterozygous. If the second allele is one of the 23 common mutations recommended in 2012 for screening by the American College of Medical Genetics and the American College of Obstetricians and Gynecologists, a sweat chloride test usually shows chloride levels that are about three-fold elevated over normal levels. Lung function is often impaired and 59 percent of these patients develop *P. aeruginosa* lung infections. About 93 percent of compound heterozygous patients also have pancreatic insufficiency.

Class III *CFTR* mutations result in defective regulation of the CFTR ion channel. These mutations usually alter the amino acids found in the NBD1 and NBD2 domains that bind ATP. ATP is the most common source of energy for use in cellular processes. It has three phosphates connected in series and the bond between the last two is particularly high in energy content. In a process called ATP hydrolysis, cleavage of the terminal bond converts ATP into adenosine diphosphate (ADP), with the release of energy. Once bound to the NBD1 or NBD2 domain, ATP is cleaved to produce ADP. Although binding of ATP to the CFTR protein is known to activate it as an ion channel, scientists do not know whether activation is triggered by the binding of ATP or by the energy of ATP hydrolysis. Class II *CFTR* mutations result in mutant proteins that either cannot bind ATP, or cannot hydrolyze ATP into ADP. Examples of this type of mutation are c.1652G>A and p.Gly551Asp, which prevent CFTR from binding ATP and opening the ion gate. As a result, this type of mutation is referred to as a gating defect mutation.

Class IV *CFTR* mutations cause a decrease in the ability of the CFTR protein to transport chloride ions across the outer cell membrane. The amino acids on the inside of the normal CFTR ion channel have chemical properties that facilitate conductance, or the flow of negatively charged ions through the channel. Class IV mutations alter the identity of an amino acid in the ion channel. For example, the mutation c.350G>A, also known as p.Arg117His, results in the substitution of an arginine for a histidine. The change in chemical properties and shape of the amino acid impair the conductance of the CFTR ion channel.

Class V mutations result in the production of normal CFTR proteins that are transported to the outer cell membrane and function properly as regulated ion channels. However, the number of CFTR proteins at the outer cell membrane is reduced by these mutations because of inefficient protein trafficking or abnormal RNA splicing. The mutation c.1364C>A, or p.Ala455Glu, is a class V mutation and a missense mutation, because the normal amino acid alanine is replaced by a glutamate. This mutation reduces the stability of the resulting CFTR protein. However, the stability is not reduced as much as it is for class II *CFTR* mutations, and so the associated phenotype is not as severe. Class V mutations frequently influence the efficiency of RNA splicing. A given class V mutation can result in the simultaneous production of normally spliced mRNA that can be translated into normal CFTR protein and abnormally spliced RNA that cannot be translated into normal CFTR protein. The ratio of normally spliced RNA to abnormally spliced RNA has been observed to vary among CF patients and even among different organs of the same patient. Class V *CFTR* splicing mutations can cause aberrant splicing that excludes part or all of an exon. Splicing mutations can also result in the abnormal inclusion of part or all of an intron.

Class VI *CFTR* mutations cause decreased stability of the CFTR protein or impair its ability to regulate other ion channels. Class VI mutations also result in a CFTR protein that spends an abnormally short amount of time at the outer cell membrane. The mutations in this class affect the cytoplasmic carboxyl terminus of the CFTR protein. Mutations that introduce a premature stop codon or cause a translational frameshift can result in the loss of 70 or more amino acids from the cytoplasmic carboxyl terminus, and the resulting protein is unstable. The phenotypic consequences of the mutations are pancreatic insufficiency, recurrent lung infections, and elevated sweat chloride. One example of a class VI mutation is c.4147_4148insA, or p.Ile1383AsnfsX3. The insertion of an extra A results in a frameshift mutation that causes the use of a premature termination codon as the third codon after the insertion.

Where Can Information about CFTR Gene Mutations Be Found?

The most comprehensive database of *CFTR* mutations is the Cystic Fibrosis Mutation Database (CFTR1), which was established by the Cystic Fibrosis Genetic Analysis Consortium in 1989 and is maintained by the Cystic Fibrosis Center at the Hospital for Sick Children in Toronto. Curated clinical information about patients carrying CFTR mutations is freely available at the Clinical and Functional Translation of CFTR (CFTR2) database, funded by the Cystic Fibrosis Foundation. There are over 2,000 mutations described in the CFTR1 database. This fact alone makes connecting *CFTR* genotype to CF phenotype a daunting prospect. Each person with mild or severe symptoms of CF has two mutant alleles. Assuming that there are exactly 2,000 mutant alleles, then there are 2,000 × 2,000, or 4 million different combinations of alleles. Although the CFTR2 database contains information about the CF symptoms of over 88,000 people from 41 counties, it does not contain nearly enough data to assign phenotypic consequences to each of the 2,000 mutant alleles. The curators of the database have simplified their site by taking advantage of high frequency of the p.Phe508del mutation. They have included clinical information about every patient included in the database who has two p.Phe508del alleles, including sweat electrolyte levels, measurement of lung function with FEV_1 percent, information about pancreatic sufficiency or insufficiency, and the likelihood of lung infection by *P. aeruginosa*. They have also included clinical information about every patient who has one p.Phe508del allele in combination with one of the 23 most common mutations detected during newborn screening. The CFTR2 database also provides clinical data from patients who have one p.Phe508del allele in combination with a second allele that is known to cause pancreatic insufficiency. With time, CFTR2 will accumulate more data that can be used to clarify the CF phenotypic consequences of an increasing number of *CFTR* allele combinations.

CF affects about 70,000 people worldwide and there are about 1,000 new cases diagnosed each year. However, the frequency distribution of *CFTR* mutant alleles is different for people of different ethnic backgrounds. The majority of people with CF have Caucasian parents. As a result, the disease is most prevalent in North America, Europe, and Australasia. However, CF occurs in nearly every ethnic group, including African, Latin American, and Middle Eastern populations. The most common *CFTR* mutation is the class II *CFTR* mutation p.Phe508del. In the CFTR2 database, 83 percent of the included CF patients have at least one copy of the p.Phe508del allele and 42 percent have two copies of it. The second most common *CFTR* mutant allele in the database is the class I mutation p.Gly542X. Five percent of CF patients in the CFTR2 database have at least one copy of the p.Gly542X allele but its frequency is elevated in some populations. For example, 25 percent of patients from the Canary Islands and 21 percent of patients from Murcia have at least one copy of p.Gly542X. The class III mutation p.Gly551Asp allele is the third most common allele in the world, and there is at least one copy of it in 4 percent of the patients in the CFTR2 database. Fourteen percent of patients from Ireland and 6 percent of patients from throughout the United Kingdom have at least one p.Gly551Asp allele. The fourth most common *CFTR* mutation is the class II mutation p.Asn1303Lys, of which there is at least one copy in 3 percent of the CF patients in the CFTR2 database. Its frequency is much higher in CF patients from Algeria, of which 20 percent have at least one copy. Ten percent of CF patients from Lebanon have at least one copy of p.Asn1303Lys. At least one copy of the class I mutation p.Trp1282X is present in 2 percent of CF patients in the CFTR2 database, making it the fifth most common *CFTR* mutation in the world. At least one copy of p.Trp1282X is present in 22 percent of CF patients from Israel and 20 percent of CF patients from Lebanon.

The p.Phe508del mutation is found in 70 percent of mutant *CFTR* alleles worldwide. However, there is a significant variation in the frequency of the p.Phe508del mutation among populations. The p.Phe508del is 30 percent of the *CFTR* mutant alleles in Turkey, 50 percent of the mutant alleles in Italy, and 88 percent of the mutant alleles in Denmark. One hypothesis to explain this difference is that migration of early farmers from the Middle East during the Neolithic period to northwest Europe resulted in a founder effect for the p.Phe508del allele. The small number of people who migrated happened by chance to carry a larger frequency of the allele than the population they left behind. Successive generations inherited the allele at that higher frequency. An alternative explanation is that the p.Phe508del allele provided a fitness advantage to people who were heterozygous. This hypothesis proposes that decreased production of the CFTR protein in heterozygotes reduced the chance for infectious pathogenic bacteria to enter the intestinal epithelium. Evidence in support of the hypothesis includes the observation that *Salmonella typhi* bacteria translocate into the intestinal epithelium at a much lower rate in mice that are heterozygous with the p.Phe508del allele compared to mice that are homozygous for the normal *CFTR* allele. No translocation of the bacteria was observed in mice that were homozygous for the p.Phe508del allele. The hypothesis of a selective advantage in people with one copy of the mutant allele is referred to as balancing selection, because there is a balance between the fitness advantage gained by heterozygotes and the fitness disadvantage experienced by people with CF. It is possible that all the p.Phe508del mutant alleles in human populations throughout the world today are descended from one original mutation event that occurred at least 10,000 years ago. Evidence for this hypothesis comes from a molecular genetic analysis of 377 mutant *CFTR* and 358 normal *CFTR* alleles. The mutant alleles had the same pattern of DNA sequences called microsatellites in their *CFTR* introns. This evidence supports the hypothesis that all of the p.Phe508del *CFTR* alleles are derived from one ancestral mutant allele.

CF Contributing Factors

CF is a Mendelian monogenic recessive disorder caused by mutations in the *CFTR* gene. Underlying this description is the hypothesis that *CFTR* genotype is the sole determinant of the phenotypic manifestations of CF disease. However, the monogenic hypothesis turns out to be much more of an exception than a rule. An example of how the hypothesis functions as a rule is that all males who carry any two *CFTR* mutant alleles that cause CF have congenital bilateral absence of the vas deferens (CBAVD). The *CFTR* genotype is 100 percent predictive of the phenotype of CBAVD. For the majority of CF symptoms, *CFTR* genotype is not the sole determinant of disease phenotype. For CF patients who have a given *CFTR* genotype, the severity and age of onset of lung disease, liver disease, pancreatic dysfunction, and CF-related diabetes vary widely. For example, it has been estimated that less than 50 percent of the variation in lung disease among CF patients can be explained by their *CFTR* alleles alone. Other contributing factors cause variation in the course of CF disease besides *CFTR* genotype. These factors can be broadly categorized as genetic or environmental, although the interplay between the two factors is complex.

Environmental factors that contribute to CF are varied. The diet of CF patients can have a major impact on the ability of the digestive system to absorb nutrients and vitamins and on its ability to process food properly. Malnutrition can affect the strength of muscles used in respiration and the health of the immune system. The severity of CF lung disease can be affected by cigarette smoking, exposure to environmental tobacco smoke, or inhalation of other particles or gases in the air that are harmful to the lungs. The climate of the region in which a CF patient lives might play a role in the development of their disease. The socioeconomic status of CF patients can also be considered an environmental factor because it influences the availability of health care. There is also research indicating that stress experienced by CF patients can affect the course of their disease.

Genetic contributing factors for CF have been termed gene modifiers of the disease. Gene modifiers vary in people without CF in ways that have no observable phenotypic consequence, but in people with CF, the variation affects CF symptoms. One approach to finding genetic modifiers is to use knowledge of the molecular and cellular basis of CF to predict genes for which genotype variation would be expected to affect the course of the disease. One advantage of this approach is that candidate genes are often chosen for which research has already been done. This means that knowledge of the effect of the gene modifier on cellular metabolism can be adapted toward an improved understanding of CF. A limitation of the candidate gene approach is that it is focused on well-understood genes and the metabolic pathways they control.

An alternative approach that allows for the discovery of new and unpredicted gene modifiers is called a **genome-wide association study** (GWAS). GWAS relies on the availability of the reference DNA sequence of the human genome, a map of human genetic variation, and methods to analyze the genomes of many study participants. Researchers isolate DNA from GWAS participants and look for variations in their genomes. People in general share about 99.7 percent of the 3.2 billion DNA bases in the human genome, which means that there are about 10 million places where the base present in one person will be different in another person. These places are called single nucleotide polymorphisms, or SNPs (pronounced "snips"). The term polymorphism means many shapes, although in this context there can only be four shapes because there are only four DNA bases. For example, it may be that there is a G present at a particular position on chromosome 5 for most people, but some people have a T in that position. To qualify as a SNP, the alternative base must be found in at least 1 percent of people. A GWAS of CF was conducted by the International CF Gene Modifier Consortium, composed of research institutions from the United States, Canada, and France. The researchers grouped 6,365 CF patients according to a phenotype score based on the FEV_1 percent measure of lung function. The researchers analyzed more than 8.5 million SNPs from the patients and looked for associations between SNPs and FEV_1 percent values. The GWAS resulted in the identification of SNPs in five regions on chromosomes 3, 5, 6, 11, and X that are associated with the severity of lung disease caused by CF. The five regions contain genes that became candidates for gene modifiers of CF lung disease. Previously determined information about the candidate genes is suggestive of the mechanisms by which they affect the pathophysiology of CF. For example, two genes in the chromosome 3 region called *MUC4* and *MUC20* encoded mucin proteins that are found on ciliated airway mucosal surfaces. Mucin genes affect the distribution of mucus in the airways. A gene on chromosome 5 called *SLC9A3* encodes a protein that is involved in the transport of ions across the membranes of epithelial cells. Variation in this gene has been implicated in the frequency of intestinal obstruction in people with CF. Other genes found in the regions encode proteins known to affect protein trafficking, proteins that contribute to the function of microtubules that form cilia on cellular surfaces, proteins that have an impact on the junctions between epithelial cells, and proteins that participate in inflammatory response of the immune system. Knowledge of the function of these gene modifiers and others can be used toward a better understanding of the molecular and cellular mechanisms by which CF disease is manifested and to develop pharmaceuticals tailored to specific gene modifier genotypes, enabling personalized treatment for CF patients.

CHAPTER 3

Treatment and Therapy

Advances in cystic fibrosis (CF) treatment and therapy have enabled CF patients to live longer and have a better quality of life. This chapter will describe the treatments and therapies available to CF patients that allow many of them to live into their 40s or 50s.

How Are CF Symptoms Treated?

Lung infections are the primary cause of pulmonary disease associated with CF, which is the leading cause of illness and death among CF patients. Chronic bacterial infection of the respiratory system and the resulting inflammation of the airways results in 80 to 95 percent of deaths due to CF. Lung infections are treated with a combination of antibiotics and medications that reduce inflammation. The most common bacteria infecting the lungs of people with CF are *Haemophilus influenzae, Staphylococcus aureus, Pseudomonas aeruginosa, Burkholderia cepacia,* and Mycobacteria species. Often, a sputum culture is used to determine which bacteria are present and, therefore, which antibiotic to administer. Antibiotics may be given orally in liquid or capsule form. Oral antibiotics are used to fight mild lung infections. A commonly used example of an oral antibiotic is ciprofloxacin, which has activity against *P. aeruginosa*, methicillin-resistant *S. aureus* (MRSA), and other pathogens. Ciprofloxacin inhibits bacterial DNA synthesis. Trimethoprim is effective against *S. aureus, H. influenzae,* and other related bacterial pathogens. It acts by disrupting bacterial metabolism associated with the production of folic acid, which is needed for DNA and amino acid synthesis. Sulfamethoxazole also inhibits bacterial folic acid production and can be used in combination with trimethoprim.

Antibiotics can also be administered intravenously with a catheter. Intravenous antibiotics are used to treat infections by bacteria that are not responsive to oral antibiotics or infections that are too advanced to be effectively treated with oral antibiotics. Tobramycin is effective against *P. aeruginosa* but cannot pass into the bloodstream from the digestive system. As a result, it must be administered by an injection into the bloodstream or into muscle. Tobramycin acts by preventing the joining of the two subunits of the ribosome, which is responsible for translation of mRNAs into proteins. Piperacillin is also effective against *P. aeruginosa*. When used with tazobactam, an inhibitor of the beta lactamase enzyme that gives antibiotic resistance to some bacteria, piperacillin is effective in controlling hospital-acquired infections. Because piperacillin cannot be absorbed through the digestive system, it is administered by intravenous or intramuscular injection.

Antibiotics may also be inhaled as an aerosol. Inhaled antibiotics deliver the medication directly to the site of lung infections. This method of delivery is used when an infection is established by bacteria that are resistant to oral and intravenous antibiotics. Antibiotics that can be aerosolized and used with inhalation include gentamicin, aztreonam, colistin, and tobramycin. For treatment of *P. aeruginosa* infection of the respiratory system, tobramycin inhalation powder, tobramycin inhalation solution, and aztreonam inhalation solution have been shown to be effective. Inhaled polymyxin is effective against lung infection by multidrug-resistant Gram-negative bacteria, including *P. aeruginosa*. Inhaled antibiotics can reduce the density of bacteria in the airways and prevent bacteria from establishing long-term colonies in the respiratory system. They are also useful in the prevention of infection after lung transplantation.

Effective treatment of bacterial infections with antibiotics is sometimes challenged by the presence of bacteria that are not susceptible to the antibiotics. For example, the lungs can become infected with antibiotic-resistant, biofilm-forming *P. aeruginosa*. It is important to determine whether *P. aeruginosa* isolated from a given patient is the type that can form biofilms or not, as this will determine which antibiotic to use. The ability of nontuberculous mycobacteria (NTM) to form biofilms also gives them resistance to antibiotics. *H. influenzae* undergoes hypermutation, which allows populations of the bacteria to generate genetic resistance to antibiotics. *B. cepacia* complex bacteria are often resistant to multiple antibiotics, although combinations of antibiotics have been used successfully. MRSA is resistant to many antibiotics, has become a major problem in hospital settings, and is found with increasing frequency in CF patients.

The response of the immune system to bacterial infection is to accumulate white blood cells called neutrophils at the site of infection that are able to ingest and destroy the bacteria. Neutrophils also release chemical messengers called cytokines that signal other cells in the immune system to begin fighting the infection. This is referred to as the inflammatory response. In lung infections as a consequence of CF, the inflammatory response leads to swelling of the tissues surrounding the airways and collateral damage to the epithelial cells that line the airways. Bronchitis, which is inflammation and swelling of the bronchi, can develop. Anti-inflammatory medications are effective in treating bronchitis. One example is prednisone, a synthetic corticosteroid hormone that suppresses the immune system and dampens the inflammatory response. Other corticosteroids such as triamcinolone and fluticasone are also used as anti-inflammatory medications. Anti-inflammatory medications that are not steroid hormones are called nonsteroidal anti-inflammatory drugs (NSAIDS). They include ibuprofen, naproxen, and piroxicam. Antihistamines such as loratadine, cetirizine, and fexofenadine are also used to treat symptoms that arise from the inflammatory response in CF patients.

Another type of medication that is used to treat respiratory symptoms in people with CF is bronchodilators. Bronchodilators help open constricted bronchi and bronchioles. One example is albuterol, which can be taken orally or with a metered-dose inhaler. Another example is theophylline, which can be taken orally or intravenously. Medications called mucolytics can be used in the treatment of CF respiratory symptoms. Mucolytics digest the thickened mucus that accumulates in the airways. The thinned and loosened mucus can then be more easily coughed up. A mucolytic called DNase functions by digesting the large DNA molecules found in the dead bacterial and human cells that accumulate in mucus. The mucolytic guaifenesin acts by increasing the volume of water in mucus, thereby reducing its viscosity. N-acetylcysteine is a mucolytic that cleaves the disulfide bonds that stabilize the structure of proteins in mucus. Mucolytics can be taken orally or with a nebulizer.

Symptoms of CF associated with the digestive system can also be treated with medications. Pancreatic insufficiency causes poor absorption of the nutrients from proteins, fats, and carbohydrates and can lead to chronic abdominal pain or severe constipation. Stool softeners may be used to treat these symptoms. Docusate is a detergent that helps to solubilize fats that accumulate in stools. Casanthranol is a type of laxative, and polyethylene glycol (PEG) helps to retain water in the stool, thereby softening it.

Treatment of some symptoms and conditions caused by CF requires medical procedures or surgery. Nasal polyps that obstruct breathing may have to be removed by a delicate surgical procedure called endoscopy, during which a camera and surgical instruments are guided into the nasal passages. Endoscopy and lavage may be used to remove mucus from obstructed airways. In cases of extreme malnutrition caused by pancreatic insufficiency, there may be a need for a feeding tube to be inserted through the nose to deliver nutrients to the stomach. Bowell surgery may be needed to remove an intestinal blockage or to repair intussusception.

CF Therapies

Although CF affects organs and systems throughout the body, its effects on the respiratory system are the most troubling for people living with the disease. The dysfunction of the CF transmembrane conductance regulator (CFTR) ion channel leads to the production of thick and sticky mucus that accumulates in the lungs, increasing their susceptibility to bacterial and fungal infections. Chronic infections damage the lungs and lead to impaired breathing. The goals of CF respiratory therapy are to minimize the risk of infections and to maintain respiratory function at the highest possible level. A key category of respiratory therapy is **airway clearance techniques** (ACTs). ACTs clear mucus from the lungs, which helps to avoid infections and the subsequent lung damage that comes from them. Many different ACTs have been developed that can be learned and performed individually by children and adults, while others require assistance or specialized equipment. A bronchodilator is often used to open the air passages before performing ACTs. Inhaled mucus thinners may also be taken before beginning ACTs. ACTs move mucus from the small bronchiole passages to the large bronchi airways, where it can be expelled by coughing. Huff coughing is an ACT that involves taking in and holding a breath for a couple of seconds and blowing it out with a slow but forceful huff of air. This is repeated two more times and followed by a strong cough to clear the mucus. The huff coughing cycle is then repeated four or five times. Chest physical therapy (CPT) uses gravity to drain mucus from each of the five lobes of the lungs. The body is positioned so that the opening of each lobe in turn is oriented downward, and clapping or vibrating of the chest is used to dislodge mucus from the lobe, allowing it to flow into the bronchi. Coughing or huff coughing can then be used to expel the mucus. The active cycle of breathing technique (ACBT) begins with breathing techniques that relax the airways. A deep breath is then taken and held while chest clapping or vibrating is used to free up mucus. Repeated huff coughing is then used to clear the mucus. ACBT takes between 10 and 30 minutes. Another type of ACT is autogenic drainage (AD). This method uses controlled breathing in three phases to separate mucus from the walls of the airways, cause it to collect in bronchi, and clear it with breathing at different speeds. AD is a learned skill that can take up to 45 minutes to complete. During an ACT called positive expiratory pressure (PEP), a mask is used that allows the free inhalation of air but restricts its exhalation. This helps air to get between mucus and the walls of the lungs and the airways. Another ACT is high-frequency chest wall oscillation, also referred to as "the vest". This involves a machine that has an air-pulse generator and an inflatable vest connected to it. The vest inflates and deflates as many as 20 times per second, sending vibrations into the chest cavity and the lungs. After about 5 minutes, the machine is stopped and huff coughing is used to clear the mucus that has been dislodged. The process is usually repeated four or five times.

Blockage of the pancreatic duct caused by CF prevents the pancreas from delivering enzymes to the small intestine that digests food. The resulting pancreatic insufficiency causes poor absorption of the nutrients from proteins, fats, and carbohydrates. This leads to poor weight gain and growth. The enzymes that the pancreas fails to provide to the small intestine in CF patients can be replaced with **pancreatic enzyme replacement therapy** (PERT). PERT may involve the use of pancreatin, which contains the enzymes amylase to digest starch, lipase to break down fats, and trypsin to digest proteins. These pancreatic enzyme formulations are taken orally right before eating. Because pancreatic insufficiency also results in a lower absorption of fat-soluble vitamins from food, CF patients often have vitamin deficiencies that can be corrected with nutritional therapy. A commonly used vitamin formulation in nutritional therapy is ADEK, named for its inclusion of vitamins A, D, E, and K. Vitamin A is needed for its role in vision, bone and tooth growth, and maintenance of a healthy immune system. Vitamin D helps to maintain the strength of bones and teeth. Vitamin E is an antioxidant that prevents damage to cells from free radicals, and vitamin K plays an important role in blood clotting. Minerals that are not properly absorbed in CF patients can be replaced as well. Calcium is needed for building bones and teeth, and for the normal function of nerve and muscle cells. Iron is required for the production of red blood cells, whereas sodium chloride maintains fluid balance in tissues throughout the body. Zinc is required for cell growth and division, wound healing, and for normal function of the immune system. Nutritional therapy is also used to maintain hydration and electrolyte levels. A diet high in calories, especially from fat, is followed to provide a lot of energy. Nutritional drinks are used to ensure absorption of nutrients. Sports drinks that contain electrolytes like sodium and potassium help to replace salts that are lost in sweat. It is usually recommended that people with CF also include table salt as a regular part of their diet, unless they also have high blood pressure. Nutritional therapy is sometimes accompanied by administration of a stool softener that prevents intestinal blockages and constipation.

Establishing and maintaining a fitness plan is also therapeutic for people with CF. Choosing a fitness plan is best done in consultation with a health care team that can evaluate the level of exercise that can be tolerated in the context of CF challenges to the respiratory and digestive systems. Regular exercise makes ACTs more effective, reduces the rate of lung infections, and improves the ability of the immune system to clear infections. Maintaining fitness also has emotional and psychological benefits for people living with CF.

CHAPTER 4

Future Prospects

Improved Diagnosis of Cystic Fibrosis

Although it is considered to be the gold standard for the diagnosis of cystic fibrosis (CF), the sweat chloride test can sometimes give equivocal results. Genetic testing can provide evidence in support of a clear diagnosis of CF, but it is challenged by incomplete information about the effect of many of the over 2,000 *CFTR* mutations described so far. Accurate diagnosis of CF is also complicated by the spectrum of symptoms that occur in individuals who have a specific *CFTR* genotype. For these reasons, there is a need for varied and improved CF diagnostic tests. One example is the transepithelial nasal potential difference (NPD) test. NPD directly measures the function of the CFTR ion channel in the cells found in the nasal epithelium. Electrodes are used to measure the electrical potential difference caused by the flow of ions across the epithelium. CF causes the electrical potential to be larger. Another CF diagnostic test is intestinal ion channel measurement (ICM). This test is performed outside the body on cells taken from a rectal biopsy. The cells are exposed to chemicals that cause them to secrete chloride ions through their CFTR ion channels while they are in a chamber that allows measurement of the rate of chloride ion secretion.

Improvements in the diagnosis of *Pseudomonas aeruginosa* infections in CF patients are also being made. *P. aeruginosa* infections are usually detected by culturing bacteria from a sputum sample or a throat swab or performing genetic analysis of the bacterial DNA. Sometimes, the results of this analysis are difficult to interpret. Proper management of an infection may also require that the results are made available faster. An alternative method detects *P. aeruginosa* bacteria that are exhaled in breath. It uses a metabolite called urea that is labeled with a harmless isotope of carbon that is heavier than normal carbon. The heavy urea is introduced into the lungs of the patient with a nebulizer. *P. aeruginosa* bacteria produce an enzyme called urease that converts urea to carbon dioxide. The heavy urea is converted into heavy carbon dioxide at a rate that depends on the number of *P. aeruginosa* bacteria in the lungs of the patient being tested. The test gives quantitative results in real time that can be used to manage *P. aeruginosa* infections.

As described in Chapter 2, genetic testing is an important component of CF diagnosis. For children or adults who are suspected of having CF because of a sweat test or lung function test, genetic testing can be used to determine which mutations of the *CFTR* gene are present. Genetic testing can also be used to determine whether someone is a carrier of a *CFTR* mutation. When a couple learns that each of them is a carrier and they decide to go ahead and conceive a child, genetic testing is particularly useful to determine if the fetus inherited the mutant *CFTR* allele from each parent and will therefore be born with CF. This is referred to as **prenatal diagnostic testing**, or prenatal screening. It involves collection of fetal cells by amniocentesis or chorionic villus sampling (CVS). Amniocentesis is normally conducted between weeks 14 and 18 of pregnancy. A needle is inserted through the abdominal wall and into the amniotic sac that surrounds the fetus. Amniotic fluid, which contains fetal cells, is withdrawn. CVS is usually done between weeks 10 and 12 of pregnancy. It involves the extraction of cells through the cervix or the abdominal wall from the chorion, which is part of the placenta. The risk of miscarriage caused by CVS is estimated to be between 0.5 and 1 percent. For *CFTR* prenatal diagnostic screening, fetal cells are used to prepare fetal DNA, and DNA testing is conducted to determine if mutations in one or both of the *CFTR* alleles are present. A positive genetic test for CF can be used by prospective parents and their health care providers to make a decision about whether to continue with the pregnancy or opt for its termination by abortion.

For couples who have ethical or moral concerns about terminating a pregnancy because of a positive prenatal diagnostic test for CF, **preimplantation genetic diagnosis** (PGD) is a possibility. PGD is carried out in the same manner as genetic testing of samples from adults, children, or fetuses, except that it is conducted on cellular samples taken from preimplantation embryos that are produced by in vitro fertilization (IVF). IVF involves the stimulation of egg maturation in the prospective mother, use of ultrasound to locate mature egg follicles in her ovaries, and harvesting of the follicles. The eggs are then mixed with semen collected from the prospective father to produce embryos that can either be implanted to establish a pregnancy or frozen for later use. Over 5 million babies have been born as a result of IVF since it was first performed in 1978. There are two methods by which a sample is extracted from IVF embryos for use in PGD. Blastomere biopsy involves allowing the fertilized egg to divide three times over the course of 3 days into an embryo that contains eight cells referred to as blastomeres. One of the eight cells is carefully removed from the embryo and used for genetic testing. At this point, all of the cells are identical and fully capable of developing into any type of cell in the body, and so the loss of one of the cells has no known effect on development. However, blastomere biopsy has been shown to reduce the frequency at which IVF embryos success-fully implant into the uterus. Because of this problem, another method was developed called trophectoderm biopsy. The embryo is allowed to develop until the fourth or fifth day after fertilization. At this point, the embryonic cells have specialized into an inner cell mass that will become the fetus and an outer layer called the trophectoderm that will join with maternal cells to form the placenta. A trophectoderm biopsy involves removal of about ten of the trophectoderm cells. This process results in a higher rate of successful embryo implantation. The results of genetic testing of fetal cells collected by one of these two methods provide genotypes for each embryo tested. Whenever possible, embryos that received two normal *CFTR* alleles are chosen to initiate a pregnancy. This has resulted in the birth of hundreds of healthy children to CF carrier parents. Embryos are destroyed prior to im-plantation if they are found by PGD to have genotypes known to cause CF. Whether or not it is ethically and morally sound to produce living embryos that have the potential to develop into human beings only to destroy them is an important but difficult question for debate by individuals, families, and societies.

Prospects for Better CF Treatment

As described in Chapter 3, most CF treatments and therapies target the symptoms of the disease. However, research has discovered new pharmaceuticals that target the root causes of the disease associated with the production, function, and regulation of the CFTR protein. One category of promising drugs is **CFTR potentiators** that allow CFTR ion channels that are present in normal numbers at the outer cell membrane, but which are defective, to open, enabling the transport of chloride ions across the membranes of epithelial cells throughout the body. The first CFTR potentiator to be approved by the U.S. Food and Drug Administration (FDA) was ivacaftor, marketed as Kalydeco by Vertex Pharmaceuticals. The FDA approval came in January 2012, and was followed by approval from the European Medicines Agency. Based on clinical trials that demonstrated improvement in lung function, weight gain, and sweat chloride levels, and a reduction in the pulmonary exacerbation rate, ivacaftor was approved for the treatment of disease in CF patients who carry the p.Gly551Asp mutation. There are about 3,000 people worldwide whose CF is caused by this mutation. This is about 4 percent of the people who have CF. Additional clinical trials resulted in approval of ivacaftor for the treatment of CF caused by eight other class III *CFTR* mutations. Ivacaftor is expensive, at a cost of over $300,000 per year (in 2016), although this is not unusual for drugs that treat rare diseases. Discussions and negotiations are ongoing between Vertex Pharmaceuticals and the governments of the United States, Canada, Australia, and several European countries that are focused on ways to make ivacaftor affordable to CF patients.

Ivacaftor was also studied in patients who are homozygous for the most common *CFTR* mutation, p.Phe508del, but no positive effects were observed. This is likely due to the fact that p.Phe508del causes abnormal folding and degradation of the CFTR protein before it can be trafficked to the outer cell membrane. Ivacaftor rescues the defective gating function of CFTR proteins that have been delivered to the outer cell membrane. The search for drugs that could address the effects of p.Phe508del led to the discovery of **CFTR correctors**. CFTR correctors are able to correct one or more of the defects found in class II *CFTR* mutations. They act by preventing CFTR proteins from being degraded, improving the movement of CFTR proteins to the outer cell membrane, or inhibiting the recycling of CFTR proteins from the outer cell membrane. A CFTR corrector candidate discovered by research on human cells in culture is lumacaftor. A clinical study has been conducted with adult CF patients who were homozygous for the p.Phe508del mutation. Although sweat chloride levels were improved by lumacaftor compared to placebo, there was no significant improvement in lung function as measured by FEV_1% or the frequency of pulmonary exacerbations.

Because both ivacaftor and lumacaftor were ineffective individually for treatment of CF caused by the p.Phe508del mutation, a study was conducted to see if they would be effective together. The hypothesis was that lumacaftor would function as a CFTR corrector to counteract the effect of the p.Phe508del mutation on CFTR folding and delivery to the outer cell membrane, and ivacaftor would act as a CFTR potentiator to allow proper CFTR ion channel gating. Two randomized double-blind, placebo-controlled clinical trials were conducted with adult CF patients who had one or two copies of the p.Phe508del mutation. The results showed that patients who were given the two drugs in combination had an improvement in FEV_1%, lower sweat chloride levels, and a reduced rate of pulmonary exacerbations compared to a placebo group. The effects of the drugs on patients who were homozygous for the p.Phe508del mutation were more significant than the effects on patients who were compound heterozygous for the p.Phe508del mutation and another mutant allele. The promising results of the trials led to the announcement in July 2015 of the approval of Orkambi, a combination of ivacaftor and lumacaftor, for use in CF patients who are at least 12 years of age and who are homozygous for the p.Phe508del mutation. Orkambi is manufactured by Vertex Pharmaceuticals. Because the cost of Orkambi is about $260,000 per year (in 2016), there is considerable discussion and debate about how best to make it affordable to patients.

Another pharmaceutical with potential to treat CF is Ataluren. There is evidence that Ataluren can correct the effects of class I mutations that result in a premature stop codon that causes the production of an incomplete CFTR protein. Because another term for a premature stop codon is premature termination codon (PTC), Ataluren is also called PTC124. Ataluren allows ribosomes to continue translating mRNA into a chain of amino acids even though it encounters a premature stop codon, especially the stop codon UGA. Ataluren is described as a CFTR production corrector or a read-through agent. Ataluren was approved in 2014 by the European Medicines Agency for use in treating patients who have Duchenne muscular dystrophy caused by a premature stop codon mutation in the *DMD* gene that codes for the dystrophin protein. A clinical trial involving CF patients with class I *CFTR* mutations showed that Ataluren resulted in 4.3 percent higher FEV_1% values and fewer pulmonary exacerbations. A drug known as VX-661 has also been shown to function as a CFTR corrector. Clinical trials indicated that it might be effective alone or in combination with ivacaftor in improving the lung function of CF patients who are either homozygous for the p.Phe508del mutation or who are compound heterozygous for p.Phe508del and p.Gly551Asp. The search for more effective CFTR correctors is ongoing.

The demonstrated positive effects of CFTR potentiators and CFTR correctors alone and in combination for CF patients with specific *CFTR* mutations are cause for optimism. The basic and clinical research approaches that led to the development of ivacaftor, lumacaftor, and Orkambi are likely to result in even more effective drugs in the future. It might also be possible to discover drugs that act as both CFTR potentiators and CFTR correctors. Some have proposed that such medications could be delivered in utero to modulate CFTR production and function to embryos or fetuses. This would have the potential of rescuing CFTR function before the onset of disease symptoms. However, there is much research to be done to demonstrate the safety and long-term effectiveness of existing and future CFTR drugs. Cost–benefit analyses will also need to be conducted to weigh the high expense of treatment against benefits to CF patients and to society. The hopeful outcome of these analyses would be financially sustainable strategies for the delivery of effective pharmaceuticals to people living with CF.

Potential for CF Gene Therapy

Wide-ranging success in the introduction of DNA into bacteria, fungi, plants, and animals for applications in biotechnology, energy, agriculture, medicine, and many other fields suggested the bold prospect of introducing curative DNA into human cells for the correction of genetic disease. The process became known as **gene therapy**. CF is a good target for gene therapy because it is caused by mutation in a single gene that has been identified, isolated, and studied. However, therapeutic effects of gene therapy require that the normal gene be delivered to the cells, tissues, and organs in that it carries out its function. Because the *CFTR* gene functions throughout the body, a comprehensive gene therapy for CF would require in vivo delivery of the gene to multiple cell types, which is beyond current technology. Because of the delivery challenge, and since lung disease is the major cause of morbidity and mortality in CF patients, gene therapy trials have focused on delivering the normal *CFTR* gene to lung cells. Accessibility to lung cells through breathing also provides a technical advantage. The first in vivo CF gene therapy trial occurred in 1993. An aerosolized adenovirus was used to target the delivery of the *CFTR* gene to the nasal and bronchial epithelium of four CF patients. Although the therapy did not cure the disease, it demonstrated the feasibility of CF gene therapy. Twenty-six CF gene therapy clinical trials have been conducted since. Many of the trials were intended to evaluate the safety of viral gene delivery methods and other gene delivery methods such as liposomes, which are lipid (fat) droplets carrying the DNA to be delivered. Although the trials have mostly focused on determining the extent to which the *CFTR* gene is delivered to epithelial cells in the airways as opposed to measuring rescued function of the CFTR ion channel or clinical benefits in terms of patient symptoms, a recent CF trial showed an improvement in lung function in patients who had received a normal copy of the *CFTR* gene delivered in a liposome by a nebulizer.

Completely successful introduction of the *CFTR* gene into lung cells for the treatment of CF will require addressing several challenges. One challenge to overcome is the natural defense system of the lungs that prevent foreign material from infiltrating them. Mucus is produced to trap bacteria, viruses, and foreign matter and cilia are used to move it away from airway surfaces. Another challenge is to determine what fraction of the lung cells must receive the normal *CFTR* gene and at what level it must be expressed for clinical benefit. It is not currently feasible to introduce the gene into 100 percent of the lung epithelial cells. It may be that improvement of symptoms will occur if the *CFTR* gene is used to produce high levels of functional CFTR protein in a small fraction of the lung cells. Alternatively, it may be that clinical benefit requires introduction of the gene into a larger fraction of the cells that use it to produce a modest level of functional CFTR protein. The effective dosage is likely to vary widely according to the particular CFTR mutations that individual CF patients possess. Another challenge for CF gene therapy research is to determine the type of lung cell that is the best target for gene delivery.

Genome Editing for CF

Gene therapy for CF would introduce a new copy of the normal *CFTR* gene that will make up for the failure of mutant *CFTR* alleles to produce enough functional CFTR protein. An alternative strategy called **genome editing** would rescue normal expression of CFTR protein by fixing the mutations in the *CFTR* alleles carried on the chromosomes carried by CF patients. A promising approach to genome editing is based on a technology called CRISPR/Cas. CRISPR stands for clustered regularly interspaced short palindromic repeats. They are repeated DNA sequences that work with CRISPR-associated (Cas) genes to give bacteria protection against foreign DNA. In 2012, Charpentier and Doudna modified the CRISPR/Cas system for genome editing. Their system uses a guide RNA to direct the Cas9 endonuclease to cut the DNA at a specific target in the genome. The break in genomic DNA is repaired by one of two mechanisms. Nonhomologous end joining can occur with the introduction of small insertions or deletions of DNA bases. The CRISPR/Cas system is useful for the introduction of DNA mutations at the genomic target. Homology-directed repair can occur if a repair template is provided. This allows the introduction of any changes to the DNA sequence, including mutation repair.

Genome editing for CF would involve the directed repair of *CFTR* mutations that cause disease. Because the gene would be expressed from its normal position in the genome, it would be under the normal regulatory control and would produce the appropriate number of CFTR protein molecules. A first step in this direction was a study that used multipotent intestinal stem cells taken from two children with CF who were homozygous for the p.Phe508del *CFTR* mutation. Multipotent refers to the ability of cells to specialize into several different but related cell types. The cells were subjected to CRISPR/Cas genome editing using homology-directed repair with a template designed to correct the mutation in the *CFTR* gene. The procedure was successful in targeting the *CFTR* gene in about half of the cells. The other half of the cells received off-target mutations in other places in the genome. The genome-edited intestinal stem cells were coaxed to develop into intestinal organoids, or miniature organs, that were observed to have normal CFTR ion channel function. This experiment is an important proof of concept for CF genome editing. CRISPR/Cas genome editing for CF and other genetic diseases is an active area of basic research that may soon result in clinical trials with CF patients.

Conclusion

Our understanding of the meaning of "The child will die whose brow tastes salty when kissed" has come far in the past four hundred years. We have learned a lot about how the same molecular and cellular dysfunction that underlies the salty kiss results in cystic fibrosis (CF) symptoms associated with the lungs and the digestive system. We know the genetic basis for CF is mutations in the *CFTR* gene, and progression of disease is influenced by other genes and environmental factors. We have learned how to diagnose CF and how to manage the disease with new treatments and therapies. We know how to conduct carrier DNA testing so that prospective parents can consider reproductive options that may avoid CF. We have developed powerful new pharmaceuticals that target the failure of the CFTR ion channel protein to regulate salt levels in mucus and sweat. We have entered an era of personalized medicine that promises treatments that can be tailored to the detailed genetic makeup of CF patients. All of these advances in knowledge and practice have translated into improved life expectancy for people with CF. Four hundred years ago, the curse of the salty kiss rang true. Babies born with CF soon died. Today, fewer than 10 percent of babies born with CF die within their first year. In 1970, children with CF lived an average of 14 years. By 1980, the average life expectancy of people with CF increased to 18 years. Today, it has risen to 37 years worldwide and has been reported to be between 42 and 50 years in developed countries. There is ample evidence that this hopeful trend will continue. CF clinical trials are underway to evaluate the effectiveness of new antibiotics and anti-inflammatory medicines for the treatment of lung infections, new mucolytic agents, and medicines that improve nutrition. The discovery of CFTR potentiators and correctors has ushered in an era of personalized medicine for CF. The approval of Orkambi for use in the treatment of CF caused by the most common *CFTR* mutation is likely to lead to even more effective drugs. Soon, a day may dawn when improvements in gene therapy and genome editing technology make the dream of curing CF come true. On that day, the curse of the salty kiss will be broken.

Bibliography

Alonso y de Los Ruyzes de Fontecha J. Diez. Previlegios para mugeres prenadas. L P Grande, Alcala de Henares, 1606, 212, 1606.

Andersen DH. Cystic fibrosis of the pancreas and its relation to celiac disease: a clinical and pathological study. Am J Dis Child 56: 344–399. 1938.

Andersen DH, Hodges RC. Celiac syndrome V. Genetics of cystic fibrosis of the pancreas with consideration of the etiology. Am J Dis Child 72: 62–80. 1946.

Brodlie M, Haq IJ, Roberts K, Elborn JS. Targeted therapies to improve CFTR function in cystic fibrosis. Genome Med 7: 101. 2015.

CFTR1. Cystic Fibrosis Mutation Database. Cystic Fibrosis Center at the Hospital for Sick Children in Toronto. 2011. (website): www.genet.sickkids.on.ca/CftrDomainPage.html?domainName=NBD1. Accessed February 29, 2016.

CFTR2. Clinical and functional translation of CFTR.2015. (website): www.cftr2.org/mutation.php?view=general&mutation_id=4. Accessed February 29, 2016.

Corvol H, Blackman SM, Boëlle P-Y, Gallins PJ, Pace RG, Stonebraker JR, et al. Genome-wide association meta-analysis identifies five modifier loci of lung disease severity in cystic fibrosis. Nat Commun 6: 8382. 2015.

Cystic Fibrosis Foundation. 2015. (website): www.cff.org/. Accessed February 1, 2016.

Mayo Clinic. Diseases and conditions: cystic fibrosis. 2016. (website): www.mayoclinic.org/diseases-conditions/cystic-fibrosis/basics/definition/con-20013731. Accessed March 18, 2016.

Ferec C, Cutting GR. Assessing the disease-liability of mutations in CFTR. Cold Spring Harb Perspect Med 2. 2012.

Kerem B, Rommens JM, Buchanan JA, Markiewicz D, Cox TK, Chakravarti A, et al. Identification of the cystic fibrosis gene: genetic analysis. Science 245: 1073–1080. 1989.

Genetics Home Reference: Cystic fibrosis. 2016. (website): https://ghr.nlm.nih.gov/condition/cystic-fibrosis. Accessed March 16, 2016.

Knowles MR, Drumm M. The influence of genetics on cystic fibrosis pheno-
types. Cold Spring Harb Perspect Med 2: a009548. 2012.

WebMD. Lung Disease & Respiratory Health Center. 2016. (website): www
.webmd.com/lung/what-is-cystic-fibrosis. Accessed February 1, 2016.

Lyczak JB, Cannon CL, Pier GB. Lung infections associated with cystic fibrosis.
Clin Microbiol Rev 15: 194–222. 2002.

Mali P, Esvelt KM, Church GM. Cas9 as a versatile tool for engineering biology.
Nat Meth 10: 957–963. 2013.

Quinton PM. Physiological basis of cystic fibrosis: a historical perspective.
Physiol Rev 79: S3–S22. 1999.

Riordan JR, Rommens JM, Kerem B, Alon N, Rozmahel R, Grzelczak Z, et al.
Identification of the cystic fibrosis gene: cloning and characterization of com-
plementary DNA. Science 245: 1066–1073. 1989.

Rommens JM, Iannuzzi MC, Kerem B, Drumm ML, Melmer G, Dean M, et al.
Identification of the cystic fibrosis gene: chromosome walking and jumping.
Science 245: 1059–1065. 1989.

Schwank G, Koo B-K, Sasselli V, Dekkers JF, Heo I, Demircan T, et al. Functional repair of CFTR by CRISPR/Cas9 in intestinal stem cell organoids of
cystic fibrosis patients. Cell Stem Cell 13: 653–658. 2013.

Tsui L-C, Dorfman R. The cystic fibrosis gene: a molecular genetic perspective.
Cold Spring Harb Perspect Med 3(2). 2013.

Welsh MJ, Smith AE. Molecular mechanisms of CFTR chloride channel dysfunction in cystic fibrosis. Cell 73: 1251–1254. 1993.

Glossary

airway clearance techniques (ACTs). Used by people with cystic fibrosis to clear mucus from their lungs, thereby reducing the risk of infections.

allele. One of several possible forms of a gene that differ in nucleotide sequence.

alveoli. Air sacs in the lungs that allow the exchange of oxygen and carbon dioxide.

autosomal recessive. A pattern of inheritance in which one form of a gene is dominant over another when both are present in the same individual.

biofilm. Formed when bacteria produce polymers that allow them to stick to each other and to a surface.

bronchiectasis. Permanent damage to the bronchi of the lungs caused by repeated infections and inflammatory immune responses.

bronchioles. The smallest branches of the lungs that connect to the alveoli, or air sacs.

carrier. A description of genotype meaning that someone is heterozygous for an autosomal recessive disease.

CFTR correctors. Drugs that are able to correct one or more of the defects found in class II CFTR mutations.

CFTR potentiators. Drugs that allow defective CFTR channels that are present in normal numbers at the outer cell membrane to open.

chronic obstructive pulmonary disease (COPD). Narrowing of the bronchi and bronchioles because of inflammation caused by chronic bacterial or fungal infections.

cor pulmonale. An enlargement of the right side of the heart that compensates for lung damage and that can lead to its complete failure.

cystic fibrosis transmembrane conductance regulator (CFTR). A protein ion channel found in the membrane of epithelial cells throughout the body whose dysfunction causes cystic fibrosis.

cystic fibrosis-associated liver disease (CFLD). Abnormal liver function caused by blockage of the bile duct by thickened mucus in people with cystic fibrosis.

cystic fibrosis-related diabetes mellitus (CFRD). Failure of the pancreas to produce and secrete insulin into the bloodstream, which is caused by chronic pancreatitis in people with cystic fibrosis.

distal intestinal obstruction syndrome (DIOS). Blockage of the intestines by undigested food and thickened mucus caused by the failure of the pancreas to produce and secrete digestive enzymes.

epithelium. Layers of cells that line the cavities and surfaces of blood vessels and form the skin.

failure to thrive. Inability to gain weight and grow, or abnormal weight loss caused by malnutrition.

gene therapy. Introduction of DNA into human cells for the correction of genetic disease.

genome editing. Methods that are able to rescue normal expression of genes by correcting mutations in the genomes of affected people.

genome-wide association study. An experimental approach that looks for correlations between DNA differences among people and disease states, which allows for the discovery of new and unpredicted gene modifiers of disease.

genotype. The forms of genes that an individual carries.

hemoptysis. Coughing up blood or bloody mucus from damage to the lungs.

human genome. Approximately 3.2 billion DNA bases found on all the nuclear chromosomes and in the mitochondria of a human being.

hyperechogenic bowel. An unusual brightness to the bowel compared to the surrounding fetal tissues and bones that can be observed during a second trimester ultrasound examination, which is caused by the abnormal accumulation of meconium in the fetal bowel.

immunoreactive trypsinogen. A precursor to the digestive enzyme trypsin that is produced by the pancreas and that may be elevated in cystic fibrosis patients.

inflammatory response. The release of chemicals from damaged or infected cells, which increases blood flow and attracts immune system cells to the site.

intussusception. Occurs when a section of the large intestine called the intussusceptum slips inside another section called the intussuscepiens and blocks the flow of digested food through the large intestine.

meconium ileus. Blockage of the ileum of the large intestine by meconium, a thick, greenish black material formed by the digestion of swallowed amniotic fluid.

mucociliary clearance. The process by which particles and pathogens are trapped by mucus, which is moved out of the respiratory system by tiny hairs called cilia on the surface of cells.

mucus. A thick mixture of antiseptic enzymes, antibodies, salts, and glycoproteins called mucins that protects epithelial surfaces throughout the body.

nasal polyps. Soft benign growths in the nasal passages or sinuses that can affect breathing, cause nasal congestion, impair the ability to smell, and cause a higher risk of bacterial infection.

neutrophils. A type of white blood cell that is able to consume bacteria during an inflammatory immune response.

newborn screening. Blood tests and other diagnostic tests that are performed in the first 2 or 3 days after birth.

pancreatic enzyme replacement therapy (PERT). Administration of enzymes for the digestion of food that are not released by the pancreas to the small intestine in cystic fibrosis patients because of pancreatic insufficiency.

pancreatic insufficiency. Abnormal function of the pancreas resulting from blockage of the pancreatic ducts that deliver digestive enzymes to the small intestine.

pancreatitis. Blockage of the pancreatic duct that leads from the pancreas to the small intestine, which results in scarring and inflammation of the pancreas.

phenotype. Traits that result from genotypes.

pneumothorax. Accumulation of air outside of the lung that causes it to become disconnected from the wall of the pleural cavity of the chest.

preimplantation genetic diagnosis (PGD). Genetic testing of cellular samples taken from embryos that are produced by in vitro fertilization (IVF).

prenatal diagnostic testing. Genetic testing of fetal cells collected by amniocentesis or chorionic villus sampling.

rectal prolapse. Condition in which part of the rectum protrudes outside the anus, caused by extreme straining.

RNA splicing. Removal of introns and splicing together of successive exons to make an mRNA that contains a contiguous protein-encoding sequence.

sinusitis. Infection and inflammation of the sinuses.

sweat test. Method for the measurement of the chloride concentration in sweat, which is a reliable diagnostic tool for cystic fibrosis.

transcription. The process by which the information in the form of the nucleotide sequence of DNA is converted into the nucleotide sequence of RNA.

translation. The process by which the nucleotide sequence of mRNA is used to direct the synthesis of proteins.

Index

Active cycle of breathing technique (ACBT), 43
Acute respiratory failure, 12
Adenosine diphosphate (ADP), 31
Adenosine triphosphate (ATP), 24, 31
Aerosol, 40
Africans, cystic fibrosis carrier, 19
Airway clearance techniques (ACTs), 42
Albuterol, 41
Alleles, 17
Allergic bronchopulmonary aspergillosis (ABPA), 13
Alveoli, 5, 12
American College of Medical Genetics, 9, 20, 30
American College of Obstetricians and Gynecologists, 9, 20, 30
Amniocentesis, 46
Antibiotics
    for bacterial infection, 7, 40–41
    for lung infections, 39–40
Antihistamines, for cystic fibrosis, 41
Asians, cystic fibrosis carrier, 19
Aspergillus fumigatus, 13
Ataluren, 49–50
Autogenic drainage (AD), 43
Autosomal recessive, 17
Aztreonam, 40

Bacterial infection
    antibiotics for, 7, 40–41
    in respiratory system, 4–7
    sputum culture for, 10
Biofilm formation, 6
Blastomere biopsy, 47
Blood test, for cystic fibrosis, 11
Bowell surgery, 42
Bronchiectasis, 12–13
Bronchitis, 5, 12, 41
Bronchodilators, for cystic fibrosis, 12, 41, 42

Burkholderia cepacia, 4, 5, 7, 10, 39

Calcium, 44
Carriers, of disease-causing allele, 18–20
Cas9, 52
Casanthranol, 42
Cetirizine, 41
CFTR1 database, 32–33
CFTR2 database, 32–33
CFTR correctors, 48–50, 55
CFTR gene, 3, 11, 20
    alleles, 34
    genotype, 35
    introduction into lung cells, 51
    mutation, 25–26, 52–53
        class I, 27, 28–30
        class II, 28, 30–31
        class III, 28, 31
        class IV, 28, 31
        class V, 28, 31–32
        class VI, 28, 32
        information about, 32–35
CFTR mRNA, 23, 27
CFTR potentiators, 48–50, 55
Chaperones, 23
Charpentier, 52
Chest physical therapy (CPT), for cystic fibrosis, 43
Chest X-ray, for cystic fibrosis, 10
Chorionic villus sampling (CVS), 46
Chronic bacterial infection, of respiratory system, 39
Chronic lung infection, 13
Chronic obstructive pulmonary disease (COPD), 11–12
Chronic pancreatitis, 14
Chronic sinusitis, 14
Cigarette smoking, 36
Ciprofloxacin, 39

Codons, 22–23
Colistin, 40
Computed tomography, 10
Congenital bilateral absence of vas
    deferens (CBAVD), 35
Cor pulmonale, 12
Cotranslational modification, 23
CRISPR/Cas system, 52
Cystic fibrosis (CF)
  affecting respiratory system, 3–7
  CFTR gene mutations
      causing, 25–32
  complications from, 11–16
  contributing factors, 35–37
  diagnosis of, 8–11, 45–47
  and digestive system, 7–8
  gene therapy, potential for, 50–52
  genome editing for, 52–53
  molecular basis of, 21–25
  pattern of inheritance for, 17–21
  sweat glands and, 1–3
  symptoms and typical age
      of onset, 2
  therapies, 42–44
  treatment of, 39–42, 47–50
Cystic fibrosis–associated liver disease
    (CFLD), 15
Cystic Fibrosis Foundation, 32
Cystic Fibrosis Genetic Analysis
    Consortium, 32
Cystic fibrosis–related diabetes
    mellitus (CFRD), 14
Cystic fibrosis transmembrane
    conductance regulator
    (CFTR), 2
  as chloride ion channel, 24–25
  dysfunction of, 4, 42
  inability of, 7
  proportion of, 25
  structure of, 23
Cytokines, 41

Diet, for cystic fibrosis, 44
Digestive system
  cystic fibrosis and, 7–8
    complications, 15
    symptoms, 1
    treatment of, 42

Disease-causing mutations, 20
Distal intestinal obstruction syndrome
    (DIOS), 15
DNA
  mutation, 26
  test for cystic fibrosis, 8
DNase, 41
Docusate, 42
Doudna, 52
Duchenne muscular dystrophy, 50

Emphysema, 12
Endoplasmic reticulum (ER), 23
Endoscopy, 42
Epithelial sodium ion channel
    (ENaC), 2
Epithelium, 1
ER-associated degradation
    (ERAD), 25
European heritage, cystic fibrosis
    carrier, 19
European Medicines Agency, 48, 50
Exercise, for cystic fibrosis, 44
Exons, 22, 32

Failure to thrive, 7
Fexofenadine, 41
Fitness plan, for cystic fibrosis, 44
Fluticasone, 41
Forced vital capacity (FVC), 10
Frameshift mutation, 29

Gallbladder, cystic fibrosis and, 14–15
Gallbladder, 15
Gallstones, 15
Gene therapy, potential for, 50–52
Genetic testing, 9, 45, 46
Genome editing, 52–53
Genome-wide association study
    (GWAS), 36
Genotype, 18
Gentamicin, 40
GoLYTELY, 15
Guaifenesin, 41

Haemophilus influenzae, 4, 5, 10,
    39–40
Heatstroke, 3

Hemoptysis, 13
Hispanics, cystic fibrosis carrier, 19
Hodges, 18
Huff coughing cycle, 43
Human genome, 22
Hyperechogenic bowel, 8
Hypertonic mucus, 4

Ibuprofen, 41
Immunoreactive trypsinogen (IRT), 8
Infertility, 16
Inflammatory response, 6
Inhaled antibiotics, for lung
        infections, 40
International CF Gene Modifier
        Consortium, 37
Intestinal ion channel measurement
        (ICM), 45
Intravenous antibiotics, for lung
        infections, 40
Introns, 22
Intussusception, 15–16
In vitro fertilization (IVF), 46–47
Iron, 44
Ivacaftor, 48–49

Large intestine, cystic fibrosis and, 14
Liver cirrhosis, 15
Liver, cystic fibrosis and, 14–15
Liver function blood tests, for cystic
        fibrosis, 11
Loratadine, 41
Lumacaftor, 49
Lungs. *See also* Respiratory system
    bacterial infection in, 5, 39–41
    damage, 12

Male infertility, 16
Meconium ileus, 8
Membrane-spanning domains
        (MSDs), 24
Mendel, Gregor, 17
Mendelian inheritance, 17–18
Minerals, 44
MUC20, 37
MUC4, 37
Mucociliary clearance, 3–4
Mucolytics, for cystic fibrosis, 41

Mucous glands, 1
Mucus, 1. *See also specific mucus*
Mycobacterium leprae, 6
Mycobacterium tuberculosis, 6

N-acetylcysteine, 41
Naproxen, 41
Nasal polyps, 13–14, 42
Natural defense system, 51
Neutral mutations, 27
Neutrophils, 5, 41
Newborn screening, cystic fibrosis
        for, 8–9
Nonsense mutation, 27, 29
Nonsteroidal anti-inflammatory drugs
        (NSAIDS), for cystic
        fibrosis, 41
Nontuberculous mycobacteria
        (NTM), 6, 40
Nucleotide-binding domains, 24, 31
Nutritional drinks, 44
Nutritional therapy, 44

Oral antibiotics, for lung
        infections, 39
Orkambi, 49
Osmosis, 24–25

Pancreas
    complications, 14
    effect of cystic fibrosis on, 8
Pancreatic duct, blockage of, 43
Pancreatic enzyme formulations,
        43–44
Pancreatic enzyme replacement
        therapy (PERT), 43
Pancreatic insufficiency, 7, 42
Pancreatin, 43
Pancreatitis, 14
Phenotype, 18
Piperacillin, 40
Piroxicam, 41
Pneumothorax, 13
Polyethylene glycol (PEG), 42
Polymorphism, 36
Polymyxin, 40
Positive expiratory pressure (PEP), 43
Prednisone, 41

Pregnancy, cystic fibrosis and, 21, 46–47
Preimplantation genetic diagnosis (PGD), 21, 46
Premature termination codon (PTC), 27, 32, 49
Prenatal diagnostic testing, 46
Prenatal screening. See Prenatal diagnostic testing
Proteasome, 25
Pseudomonas aeruginosa, 5, 6, 31, 33, 39–40, 45–46
PTC124. See Ataluren

Rectal prolapse, 8
Residual volume (RV), 10
Respiratory system
    chronic bacterial infection of, 39
    cystic fibrosis and, 3–7
        complications, 11–14
        symptoms, 1
    treatment of, 41
    upper. See Upper respiratory system
RNA splicing, 22

Salmonella typhi, 34
Salty mucus, 4
Secretory glands, 1. *See also specific glands*
Single nucleotide polymorphisms (SNPs), 36
Sinusitis, 4, 14
Sinus mucus membranes, 3–4
Sinus X-ray, for cystic fibrosis, 10
SLC9A3, 37
Small intestine, cystic fibrosis and, 14
Sodium chloride reabsorption, 2–3
Splicing mutations, 29
Sports drinks, 44
Sputum culture, for bacterial infections, 10, 39
Staphylococcus aureus, 5, 39–40

Stool softeners, 42
Submucosal glands, 4
Sulfamethoxazole, 39
Sweat chloride test, 45
Sweat, definition of, 2
Sweat glands, cystic fibrosis and, 1–3
Sweat test, 9

Tazobactam, 40
Theophylline, 41
Thrive, failure to, 7
Titin gene, 22
Tobacco use. *See* Cigarette smoking
Tobramycin, 40
Total lung capacity (TLC), 10
Transcription, 22
Transepithelial nasal potential difference (NPD) test, 45
Translation process, 22–23
Triamcinolone, 41
Trimethoprim, 39
Trophectoderm biopsy, 47
Trypsinogen, 8

Upper respiratory system
    bacterial infection in, 4
    complications of cystic fibrosis affecting, 13–14
U.S. Food and Drug Administration (FDA), 48

Vitamin A, 44
Vitamin D, 44
Vitamin E, 44
Vitamin formulation, in nutritional therapy, 44
Vitamin K, 44
VX-661, 50

Zinc, 44

FORTHCOMING TITLES IN OUR HUMAN DISEASES
AND CONDITIONS COLLECTION

A. Malcolm Campbell, *Editor*

- *Gradual Loss of Mental Capacity from Alzheimer's, Volume I* by Mary Miller
- *Gradual Loss of Mental Capacity from Alzheimer's, Volume II* by Mary Miller
- *Auto-Immunity Attacks the Body* by Mary Miller
- *Uncontrolled Bleeding from Hemophilia* by Todd Eckdahl
- *Brain Degeneration from Huntington's Disease* by Todd Eckdahl

Momentum Press offers over 30 collections including Aerospace, Biomedical, Civil, Environmental, Nanomaterials, Geotechnical, and many others. We are a leading book publisher in the field of engineering, mathematics, health, and applied sciences.

Momentum Press is actively seeking collection editors as well as authors. For more information about becoming an MP author or collection editor, please visit http://www.momentumpress.net/contact

Announcing Digital Content Crafted by Librarians

Concise e-books business students need for classroom and research

Momentum Press offers digital content as authoritative treatments of advanced engineering topics by leaders in their field. Hosted on ebrary, MP provides practitioners, researchers, faculty, and students in engineering, science, and industry with innovative electronic content in sensors and controls engineering, advanced energy engineering, manufacturing, and materials science.

Momentum Press offers library-friendly terms:

- perpetual access for a one-time fee
- no subscriptions or access fees required
- unlimited concurrent usage permitted
- downloadable PDFs provided
- free MARC records included
- free trials

The **Momentum Press** digital library is very affordable, with no obligation to buy in future years.

For more information, please visit **www.momentumpress.net/library** or to set up a trial in the US, please contact **mpsales@globalepress.com**.

CPSIA information can be obtained
at www.ICGtesting.com
Printed in the USA
FFOW01n1935171217
44079170-43345FF